FAO
FISHERIES
TECHNICAL
PAPER

426/1

Management, co-management or no management?

Major dilemmas in southern African freshwater fisheries
1. Synthesis report

by
Eyolf Jul-Larsen
Jeppe Kolding
Ragnhild Overå
Jesper Raakjær Nielsen
Paul A.M. van Zwieten

FOOD AND AGRICULTURE ORGANIZATION OF THE UNITED NATIONS
Rome, 2003

The designations employed and the presentation of material
in this information product do not imply the expression of any
opinion whatsoever on the part of the Food and Agriculture
Organization of the United Nations concerning the legal or
development status of any country, territory, city or area or of
its authorities, or concerning the delimitation of its frontiers or
boundaries.

The views expressed in this publication are those of the author(s) and do
not necessarily reflect the views of the Food and Agriculture Organization
of the United Nations.

ISBN 92-5-104919-X

All rights reserved. Reproduction and dissemination of material in this
information product for educational or other non-commercial purposes are
authorized without any prior written permission from the copyright holders
provided the source is fully acknowledged. Reproduction of material in this
information product for resale or other commercial purposes is prohibited
without written permission of the copyright holders. Applications for such
permission should be addressed to the Chief, Publishing Management
Service, Information Division, FAO, Viale delle Terme di Caracalla, 00100
Rome, Italy or by e-mail to copyright@fao.org

© FAO 2003

PREPARATION OF THIS DOCUMENT

The present report is the main result of a four years research project on freshwater fisheries development in the South Africa Development Community (SADC) area funded by the Norwegian Research Council. It has involved a number of African and European researchers who all have delivered written contributions. The report is divided into a syntheses part and ten case studies covering five important freshwater bodies in the Democratic Republic of the Congo, Malawi, Zambia and Zimbabwe. Due to practical and financial constraints, it was unfortunately not possible to include all participants in the development of the synthesis but we hope that we have been able to reflect all the major findings which emerged from the case studies. The names of the main authors appear in alphabetical order.

ACKNOWLEDGEMENTS

First of all we wish to thank the Norwegian research council as the main funding sources; they have shown great interest in our work and been very supportive. The fisheries authorities in Malawi, Zambia and Zimbabwe have provided very useful assistance by giving us access to all sorts of data and we thank them sincerely. We also wish to thank the Development Planning Service, Fisheries Policy and Planning Division of the Fisheries Department of FAO, for a close and very fruitful collaboration which has included several workshops and seminars and a six months stay as visiting scientist. We also thank the Norwegian Agency for International Development (NORAD) for financial support to the dissemination of results.

Eyolf Jul-Larsen
Project Coordinator

The authors

Eyolf Jul-Larsen, Social Anthropologist at the Chr. Michelsen Institiute, Bergen, Norway, (Eyolf.Jul-Larsen@cmi.no)
Jeppe Kolding, Fisheries Biologist, Dept. for Fishery and Marine Biology, University of Bergen, Norway, (Jeppe.Kolding@ifm.uib.no)
Ragnhild Overå, Cultural Geographer, Chr. Michelsen Institute, Bergen, Norway, (Ragnhild.Overaa@cmi.no)
Jesper Raakjær Nielsen, Economist, Institute for Fisheries Management, Hirtshals, Denmark, (jrn@ifm.dk)
Paul A. M. van Zwieten, Fisheries Biologist, Fish Culture and Fisheries group, Dept. of Animal Sciences, Wageningen University, The Netherlands, (Paul.vanZwieten@Alg.VenV.wag-ur.nl)

Distribution:

FAO Members
Other interested nations and national and international organizations
FAO Fisheries Department
FAO Regional and Subregional offices

Jul-Larsen, E.; Kolding, J.; Overå, R.; Raakjær Nielsen, J.; Zwieten, P.A.M. van.
Management, co-management or no management? Major dilemmas in southern African freshwater fisheries. 1. Synthesis report.
FAO Fisheries Technical Paper. No. 426/1. Rome, FAO. 2003. 127p.

ABSTRACT

This report synthesizes the findings of ten case studies published in FAO Fisheries Technical Paper 426/2. The case studies have been conducted in five medium sized lakes in the Democratic Republic of the Congo, Malawi, Zambia and Zimbabwe. The synthesis focuses on three features relevant for the management of freshwater fisheries in the South Africa Development Community (SADC) region:

- How has fishing effort developed in these lakes over the last 50 years?

Despite a considerable increase in the total fishing effort in the region, the report demonstrates great variation in effort dynamics both in time and place. The report distinguishes between changes related to the number of people and to changes in technology and investment patterns and shows that most of the increases in effort have been population-driven rather than investment-driven.

- What causes the changes in fishing effort?

The level of mobility among fishermen - into as well as out of the fisheries sector – is considerable and this mobility is strongly influenced by economic features external to the sector (such as changes in the Copperbelt economy). Changes in the number of fishermen also depend on the effectiveness of the local access regulating mechanisms found to exist in all the lakes. The moderate prevalence of investment-driven changes in these fisheries is analysed with reference to deficiencies in infrastructure, credit support and complex and often unclear social relations prevailing at the local level. When occurring, investment-driven increases are generally induced by access to external financial sources.

- How do fishing effort and environmental factors compare in their effects on the regeneration of fish stocks?

In the five lakes studied, environmental drivers are often more significant than fishing effort in explaining changes in fish production and the strong environmental influence is not only restricted to cases where environmental variability is very high (e.g. Lake Chilwa). Total yields in the multi-species and multi-gear fisheries are surprisingly stable over a large range of effort levels, but changes in species and size composition are considerable. So, in these fisheries with small-scale operations there is limited danger in increased diversification of fishing patterns and they are close to an overall unselective and ecologically sound fishing pattern, highly adaptive to changing conditions. The danger for fish stocks increases with increased gear efficiency.

Fisheries in the SADC freshwaters are found to function as an economic buffer and as a safety valve for thousands of people moving in and out of the fisheries according to the opportunities in the national economies. At the same time the stocks tend to be less threatened than many tend to believe. Classical management theory's emphasis on limiting numbers of fishermen and co-management strategies such as exclusive economic zoning may represent a danger to the stability of this situation, even where management may be required to maintain biodiversity. There may be a need also to monitor and establish measures to control investment-driven growth in effort.

CONTENTS

Page

1. SUSTAINABLE EXPLOITATION AND SENSIBLE MANAGEMENT OF SOUTH AFRICA DEVELOPMENT COMMUNITY (SADC) FRESHWATER FISHERIES — 1

    1.1 Introduction — 1
    1.2 Our point of departure – some uncomfortable feelings — 1
    1.3 Research questions and organization of the report — 7
    1.4 Methods and data — 7

2. CONCEPTUAL CLARIFICATIONS — 9

    2.1 Fisheries concepts — 9

        2.1.1 Fishing effort, catch and catch rate — 9
        2.1.2 Catchability — 11
        2.1.3 Selectivity — 12

    2.2 Social and economic concepts — 14

        2.2.1 Population- and investment-driven changes in fishing effort and demographic change — 14
        2.2.2 Social institutions — 15

3. TRENDS IN YIELDS AND FISHING EFFORT OVER THE LAST 50 YEARS — 16

    3.1 Trends in yields and effort in the SADC region — 17

        3.1.1 Trends in yields — 17
        3.1.2 Trends in effort development — 18

    3.2 Three lakes: different histories – different effort trajectories — 21

        3.2.1 Lake Mweru — 21
        3.2.2 Lake Kariba (Zambia) — 23
        3.2.3 Lake Malombe and the South East Arm of Lake Malawi — 25

    3.3 Trends in effort dynamics: Population-driven rather than investment-driven growth — 28
    3.4 Mobility among fishermen prevents fishing from becoming a 'last resort' — 31
    3.5 Effort development in the context of the fisheries management history — 33

4. FACTORS BEHIND CHANGES IN FISHING EFFORT — 35

    4.1 Introduction — 35
    4.2 Factors influencing population-driven growth in effort — 37

        4.2.1 Natural productivity and macroeconomic changes — 37
        4.2.2 Local access-regulating mechanisms — 39
        4.2.3 Capital investments in the fisheries — 43

    4.3 Factors impeding investment-driven growth — 44

        4.3.1 Market development — 45
        4.3.2 Infrastructure — 45

			Page
	4.3.3	Production-distribution linkages	46
	4.3.4	Capital	47
	4.3.5	The local institutional landscape	47

5. THE EFFECTS OF FISHING AND ENVIRONMENTAL VARIATION ON THE REGENERATION OF FISH STOCKS — 50

    5.1 Introduction — 50
    5.2 Classical approaches — 51
    5.3 System variability: water level as environmental driver — 58
    5.4 Susceptibility of fish stocks and species to fishing under environmental variation — 66
    5.5 Selectivity and scale of operation of fishing patterns: dynamics of fishing effort — 75
    5.6 Information on catches, effort and environment in African freshwaters — 81

6. IMPLICATIONS FOR FISHERIES MANAGEMENT — 83

    6.1 Summary of findings — 83
    6.2 What are the implications for management from a biological perspective? — 84
    6.3 Social findings and management implications — 88
    6.4 Management, co-management or no management in southern African freshwater fisheries? — 91

7. REFERENCES — 95

Appendix 1   THEORETICAL ELABORATION ON ECOLOGICAL CONCEPTS AND THE DEVELOPMENT OF FISHING PATTERNS IN MULTISPECIES FISHERIES — 117

Map of study area

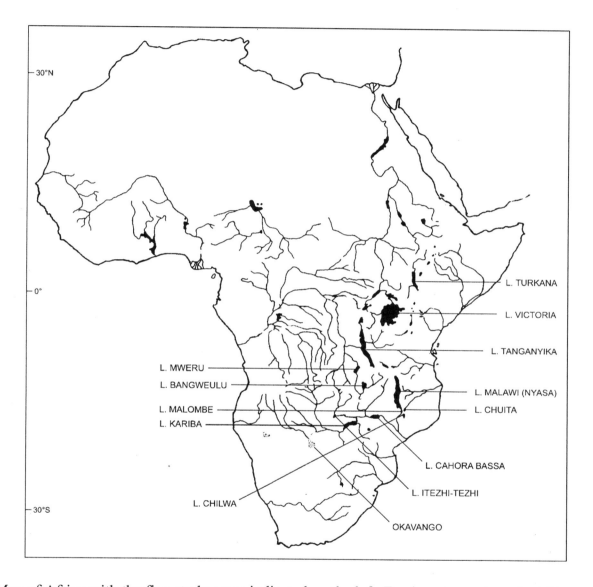

Map of Africa with the five study areas indicated on the left. Fresh water systems in Africa where the authors have additional research experience are indicated to right. (Drawn by Elin Holm).

1. SUSTAINABLE EXPLOITATION AND SENSIBLE MANAGEMENT OF SOUTH AFRICA DEVELOPMENT COMMUNITY (SADC) FRESHWATER FISHERIES

1.1 Introduction

From the hilltop of Kariba town in Zimbabwe, the spectacular view of Lake Kariba is instructive for someone concerned with fisheries management. An overview of the lake's eastern basin, from the Bumi Hills in the west, towards the Sanyati and Gatchegatche estuaries further east and all the way around to the Kariba dam in the north, reveals that the space allocated to fisheries along this lakeshore is very limited. Most of the shore is allocated for other purposes, such as natural reserves and parks, tourism, urban development and generation of hydroelectric power. A large portion of "state land", however, is not reserved for any particular purpose. The areas permanently allocated for fishing are divided between the premises of the offshore Kapenta fishery (mainly operated by white entrepreneurs) and limited areas of communal lands where other people are allowed to fish. Approximately 50 inshore fishermen organized in cooperatives (of whom many are foreigners from Malawi, Mozambique and Zambia easily expelled by a simple political decision) fish there. For the Kapenta fishing units, the spatial limitations caused by the land allocation system is a minor problem, since their modern fishing rigs can be moved to all areas of interest in the basin. The inshore fishermen, on the other hand, who are much poorer and operate from non-motorized canoes and small boats, are effectively prevented from harvesting many of the most productive fishing grounds in the basin. Their fishing effort is therefore spatially concentrated in small areas where fish become scarce and profits are consequently low.

Looking at Lake Kariba over time, one notices that the lake acts like a "living organism". Water levels and lakeshore formations constantly change – seasonally and annually. During the latter half of the 1990s, for example, the maximum lake level increased by 11 m and brought the lake back to the size and shape it had in 1963 when it was filled up for the first time.

This complex and variable situation in both time and space – sensed from the hilltop, but concretely experienced by the fishermen every day – reminds us of issues that tend to be forgotten in the literature and debates on fisheries and fisheries management. Firstly, fish is often only one of many resources in a freshwater ecosystem. The management of fish resources is therefore also affected by how the other resources are exploited. Secondly, the situation described above shows that it is the small-scale fishermen that are most negatively affected by existing resource management measures: compared with the Kapenta operators, the tourist agents, hotel managers and electricity suppliers, the poor farmers and fishermen in the communal areas around Lake Kariba have few chances (if any) of making their voices heard. In other words, fisheries management is (and must be) politics: it regulates and differentiates between people's and groups' access to vital resources. Hence the management system in Lake Kariba thus inevitably reflects local as well as national relations of power.

1.2 Our point of departure – some uncomfortable feelings

Over the years, the notions of "management" and "co-management" have come to carry considerable symbolic content. Sometimes these notions are perceived as "metaphors" representing different schools of thought in the governing of the world's renewable natural resources. Since many of these resources are not privately owned, "management" usually refers to ideas emerging from the classical work generally known as common property theory (CPT), as this evolved on the basis of models developed by M.B. Schaefer (1954), H.S. Gordon (1954) and later G. Hardin (1968). Through the development of concepts like carrying capacity, maximum sustainable yield (MSY), resource rent and maximum economic yield (MEY), we are all familiar with how adherents of CPT argue that, unless an external agent intervenes and regulates access to and exploitation of common resources, the contradiction between commonly owned resources and privately owned means of production will lead to intensified exploitation beyond nature's carrying capacity. The ultimate result is a tragedy characterized by serious reductions in the resource base (including both biomass and biodiversity) and the depletion of potential economic profits that could have been generated. The emphasis on external resource regulation has also led to an image of the "management" protagonists as particularly strong supporters of the role of the state in resource management, even if CPT as such does not automatically imply this viewpoint.

Unlike the "management" position, which became extremely influential in both academic and policy-oriented debates, the "co-management" approach is not based on one particular theory or model. In many ways it emerged as the result of different arguments contradicting the "management" view. Already in the 1970s – long before anyone had heard about the notion of "co-management" – social anthropologists and human ecologists (e.g. J.M. Acheson, B. McCay and F. Berkes) had started to challenge some of the assumptions underlying CPT. These scholars based their criticism on empirical evidence from various types of production systems and from many parts of the world. They aimed mainly to demonstrate that many production systems contain mechanisms that effectively regulate people's access to common resources. They argued that, by taking for granted that a common property regime implies free access to the resources, CPT fails to analyse and understand the social institutions that are often crucial to the management of resources in local communities. Knowledge of how these institutions work is, according to these scholars, a prerequisite for the improvement of local resource management systems. The emphasis on local management regimes as "community-based" represented an alternative to the state-centred solutions advocated by the protagonists of CPT.

In the 1980s, management studies were influenced by a new interest in institutional analysis, particularly among political scientists like E. Ostrom. Though these scholars were inspired by anthropological studies of local communities' rules for the management of common resources, the "new institutionalism" viewed natural resource management from a slightly different angle. Instead of the strong emphasis on local communities in the management of natural resources that many anthropologists had advocated, management became more of a question to be resolved in the interface between the state, the civil society and the market. The term "co-management" thus slowly came to substitute that of "community-based" management, even if the critique of and the disagreements with CPT thinking in certain ways remained the same.

Although we have chosen "Management, co-management, or no management?" as the title of our work, our aim is not to enter into the debate between the "managers" on the one side and the "co-managers/community-based managers" on the other. Instead, our approach emerges as a consequence of some uncomfortable feelings connected with this debate. Through our experience in research and other work related to African fisheries, we have increasingly realized the extent to which the issues and problems that emerge (to us) as the most important ones, are *not* addressed in the management debate. Sometimes contradictions are "produced" in the debate that do not result in particularly useful discussions and, in other instances, the debate is often incapable of establishing alternative points of view, which in our view are highly needed. Let us outline three reasons that to some degree explain why we do not find the current "management" versus "co-management" controversy particularly fruitful.

1. The "truth" problem. In our view, the debate on the management of common property resources has to a large extent centred on whether CPT as a model is "true" or not. For example, many of the intensive and detailed empirical analyses of how local institutions regulate people's access to the resources have explicitly aimed to show one of the main arguments of CPT about the need for state intervention to be false. The contributions of the community-based and co-management literature have undoubtedly been important. In a field which 20 years ago still remained completely dominated by natural sciences, they correctly make the point that CPT as a model lacks any reference to social and political content in the relations between the social actors and therefore that CPT fails to capture the fact that communities in 'real life' often develop precautions and limitations through the establishment and maintenance of a range of social institutions or "unwritten laws".

Nevertheless, we agree with those who argue that the debate has been complicated by the general drive to show CPT to be false (Brox 1990). Briefly put, Brox's argument is that much of the critique is confused because CPT is perceived (by its defenders as well as by those who criticize CPT) as an *empirical model*, in other words a model that pretends to say something about reality. However, since the model is void of social content it can only be regarded as an *analytical model* – i.e. a model that claims to have empirical relevance only if the assumptions on which it is based are correct. The question in this work is therefore *not* whether common property theory is "true" or "false", but rather whether and to what extent CPT can be seen as a useful tool in explaining the way in which SADC freshwater fisheries are developing. And this leads us to the second problem.

2. The "effort" problem. Since both 'managers' and 'co-managers' tend to consider CPT as an empirical model, few of them have seemed particularly interested in empirical investigations about the dynamics of fishing effort, or about its causes and its consequences. Like the management supporters, the protagonists of co-management accept CPT's assumption that growth of effort in a commons is inevitable, unless effective management measures are put in place. Our own experience from research and commissioned studies in many areas in Africa (marine as well as freshwater) has been far more varied. While we felt that certain fisheries in Africa arguably have been growing over a long time, it seemed that others have experienced a much more static and less dramatic development. In the latter case we also had problems identifying any effective management measures in place – initiated by the state or by the local community – and we wanted to investigate in some detail what had been the case in SADC freshwater bodies.

The causes for increased fishing effort in specific areas are seldom investigated, but merely reflect the simplistic economic and demographic arguments underlying CPT. Fisheries are seen as closed systems and so it is not envisaged that people may adapt in unexpected ways to changing circumstances, and that the extent to which they will overfish often depends on whether they have other and better economic options. Approaches such as the 'last resort' perspective of D. Pauly do consider fisheries as part of a wider social and economic context. However, whereas growth in effort is seen as a result of artisanal fisheries' ability to absorb superfluous labour from other sectors, 'last resort' thinking tends to assume that once marginalization has forced people into fisheries they are forced to remain there until they have undermined their own livelihoods. Empirical evidence, however, shows that poor people in variable and fragile environments tend to diversify their sources of income and that fisheries is often only one of several income-generating activities, or a temporary option.

Furthermore, it is interesting to notice how the co-management approach does not always incorporate important findings of the new institutional researchers. For example, the economist J.-P. Platteau (1989a and b) has developed an argument on how high 'transaction costs' in many developing countries may explain why small-scale fisheries, based on a simple technology, often seem to prevail despite the emergence of new and more capital intensive fisheries. Even though the argument has clear implications for effort development compared to the underlying assumptions of CPT, it has not led to revisions among the supporters of co-management about how they conceive effort growth.

3. The "management belief" problem. In the management/co-management controversy, the focus has often been more on *who should* manage than on *what should* be managed. Whether the enforcement of management rules is put in the hands of the state, or of the local community, or one promotes a cooperation between these two levels in the management process, both approaches are based on the assumption that fisheries can always be fruitfully managed. However, from both a natural science and a social science perspective, the truth of this assumption is not evident. New literature in both fields indicates a much more complex situation and reference needs to be made to at least three types of interrelated scientific discourses where such problems are discussed.

The first discourse relates to what has been termed the *'new ecology'*. During the 1990s, a range of detailed empirical studies representing an alternative approach in ecological research emerged. Interestingly, while both conventional management thinking and the common property debate have been richly inspired by research in fisheries, this has not been the case in the new ecology, where biological and social assumptions in conventional management thinking are being re-questioned as an entity. The new ecology discourse has mainly derived from studies of pastoralism (Ellis and Swift, 1988; Scoones, 1995; Sanford, 1983; and Swift, 1988) forestry (Fairhead and Leach, 1994, Cline-Cole, 1994) and other types of land use (Adams, 1992; Scoones and Cousins, 1994; Tiffen *et al.*, 1994). This does not mean that similar positions in biology and in social science are not found within the domain of fisheries research. During the course of this work it was certainly astonishing to come across the Memorandum of Dr C. F. Hickling[1] (see chapter 3), a fisheries biologist in the Colonial Office in London, who already in 1952 emphasized many of the same viewpoints as one finds with the new ecologists. But in more recent debates on fisheries management, very few works have tried to relate to the arguments. One notable exception is J. Quensière and others' study of the Niger's Central delta in Mali (Quensière, 1994).

Representatives of the new ecology question the long-reigning image of considering ecosystems (without human intervention) as closed entities in equilibrium (or in a process towards re-establishing equilibrium). Instead they insist that local ecosystems in most cases must be considered to be in a constant and ever-changing state of non-equilibrium due to considerable climatic variation (rainfall, temperature, wind and evaporation) which are variables defined as external to the system.

---

[1] Zambia National Archives Reference No: Sec 6/570. C.F. Hickling, "Memorandum on fisheries legislation", 4th November 1952.

The protagonists of the new ecology are concerned with the implications of these views for how one conceives the relationship between human behaviour and the regenerating ability of ecosystems. Management models based upon CPT represent an equilibrium approach and consider human intervention as the only significant external variable. This makes it possible to establish a very close correlation between human intervention (effort) and the generating capacity of the system. For example, Schaefer's (1954) dome-shaped model of the relation between effort and production and the concept of 'maximum sustainable yield' are in fact based upon the assumption that such a correlation exists. However, if the other variables are not considered as only minor 'disturbances', but as factors that may alter the dynamics of the ecosystem in a significant and perpetual manner and keep it in a state of non-equilibrium, the picture becomes far more complicated. Then one cannot *a priori* say how changes in effort will affect the ecosystem, since the effect of effort must be expected to vary according the state of the abiotic variables. An example of this line of reasoning is Ellis and Swift's (1988) demonstration of how increased effort among Kenyan pastoralists during times of abundant pastures (due to favourable rainfall) is a sound adaptation since the grass that is not utilized must be expected to die when the good rainfalls cease. There is therefore not much rationale in trying to regulate effort under these conditions.

Secondly, comes the important debate about the politics of resource conservation. Even if it may appear as a truism, it is important to remember that management and conservation measures always reflect political choices. As mentioned above, the rules of land tenure along the Zimbabwean shores of Lake Kariba affect various groups of fishermen and users of the lake differently. Studies of colonial and post-colonial conservation policies (Anderson and Grove, 1987; Beinart, 1989; Carruthers, 1993; Hoben, 1995) have demonstrated how these policies were basically rooted in the prevailing images about man and nature, of the colonial administrators: "…the function of Africa as a wilderness in which European man sought to rediscover a lost harmony with nature and the natural environment"; "This perceptual polarisation of a despoiled Europe and a 'natural' Africa has held sway since the nineteenth century"; "The problem is rooted in the nature of the colonial relationship itself, which allowed the Europeans to impose their image of Africa upon the reality of the African landscape" (Anderson and Grove, 1987: 4-5).

At the same time the studies show how regulation and conservation policies, also directly reflect the economic and political interests of the colonial state and/or the dominant groups of people influencing its decisions (e.g. white settlers). However, this does not mean that conservation policies were the same as conservation practices. Various local groups soon established patterns of behaviour that became crucial to practice even if they were ignored as policies (McCracken, 1987). Interestingly, it seems as if the post-colonial states to a large extent have prolonged many of the regulations introduced during colonial times, irrespective of changes in policy (Malasha, 2003).

Thirdly, a discourse in anthropology and sociology, with reference to sub-Saharan Africa, questions the concept of 'institution' and the distinction between so called 'modern' and 'traditional' institutions among the new institutionalists (e.g. Berry, 1993, 1997; Peters, 2000). These studies have shown that the ways access to vital resources are obtained and secured, are far more complex than the new institutionalist co-managers tend to think. By conceiving institutions as simple 'rules of the game', the new institutionalists tend to emphasize the existence in the local community of existing rules (traditional) which often oppose the formalized rules of the state (modern). The works of, for instance, Berry (1989, 1993), and

Peters (2000), show how the new institutionalists fail to grasp the historical dimension in how rules have been constructed and they convincingly argue that local institutions, crucial in the regulation of people's access to vital resources, often reflect a number of different meanings, are unclear and sometimes even lack coherence. Through detailed historical analyses, their research shows that so-called 'local management systems' have often emerged as a result of negotiation and accommodation in long-term power struggles. Various actors including a variety of external powers operating at different periods of time have based their claims and legitimacy on different 'logics' and values.

The lack of hegemonic power at the local level has facilitated a continued co-existence of different repertoires of values. When people seek to legitimate their claims to resources, this situation gives them a possibility of referring to a variety of logics which easily leads to the introduction of new ones without replacing those already existing (Berry, 1989). The neo-institutional image of a clear-cut distinction between modern and traditional institutions is challenged by the demonstration that history has entailed a much more complex institutional landscape where both the local as well as government institutions are continuously being constructed in the interface between different actors and where ambiguity and sometimes even contradiction are common. This challenge of the neo-institutional approach is therefore often referred to as a 'constructivist' approach.

These discourses have strong implications for how to legitimize, conceive and implement resource management in rural Africa. The new ecology's emphasis on non-equilibrium in ecological processes and on the role of the abiotic variables for regeneration of stocks, tends to reduce the biological rationale for a great number of detailed regulations which have been introduced into the African commons over more than a century. As shown by Ellis and Swift (1988), it may sometimes be futile to reduce pressure on pastures because the vegetation –if not eaten – will die anyhow. However, this does not mean that the new ecology considers all management measures as biologically futile. The question is merely to distinguish measures based on empirical observations and those based on dubious assumptions. Policy analyses of natural resource management sometimes question both its biological and its social legitimacy. When interests of power become one of the basic principles for regulating resource access, it is not evident that these principles reflect either ecological or social equity concerns.

Finally, the anthropological view on institutional change in Africa renders the conception of 'how to manage' or 'who should manage' much more complicated. According to this approach, it is not possible – as the neo-institutionalists would do – to distinguish between a set of modern and a set of traditional institutions that are more or less internally coherent. The concept of the 'traditional management system' – utilized extensively in the community-based management and co-management literature – becomes problematic because it is more a question of many *access-regulating mechanisms* that often are based on very different systems of norms and values and therefore may be both ambiguous and contradictory (Fay, 1994; Jul-Larsen and Kassibo, 2001). To implement management in the classical sense of a top-down set of government regulations or co-management in the sense of a co-existence between two sets of institutions (modern and traditional) therefore becomes difficult.

1.3 Research questions and the organization of the report

The preceding section has revealed a number of empirical and theoretical problems that are seldom dealt with in a systematic manner in applied research of this kind. It is therefore the intention of this work to analyse the following questions in some detail:

- How have yields and effort developed in SADC freshwater bodies?
- What have the causes behind this development been?
- How does increased fishing effort affect the regeneration of stocks?
- What are the management implications?

The objective is not so much to provide clear answers to all these questions, as to establish new bases of knowledge for the debates and indicate directions for further investigations. As the subtitle indicates, resource management is always embedded in dilemmas to which there are no simple answers. The style may therefore sometimes appear as more of a discussion than a presentation of clear-cut arguments and conclusions.

After some definitions and conceptual clarifications in the next chapter, Chapter 3 addresses 'the effort problem'. Instead of assuming that fishing effort is continuously growing, the chapter investigates the effort dynamics and to a minor extent the yield dynamics over the last 50 years, on both an aggregate level as well as on selected water bodies. In order to analyse the causes behind the development in fishing effort, the chapter distinguishes between effort growth caused by increased numbers of people (population-driven) and growth caused by increased financial investments (investment-driven). Furthermore, it investigates the occupational mobility of fishers, both in terms of getting access to the sector, and leaving it. Finally we look at the role government regulation seems to have played in explaining the effort development.

Chapter 4 is a more detailed analysis of the causes behind the changes in effort we have observed and gives us an opportunity to discuss in some detail the relevance of the new institutionalist and the constructivist approach with reference to the 'management belief problem'. The first section deals with causes behind population-driven changes, while the second investigates the requirements for investment-driven changes. Chapter 5 first discusses the results and value of classical stock-assessments, where effort is the only explanatory variable, in the freshwater systems studied. Then, taking into account the often overlooked aspects of environmentally driven natural variability in fish stocks, it examines the evidence for the long-term effects of increased effort. The impact of fishing on ecology and stock regeneration is addressed step by step from three perspectives: *system variability, susceptibility of fish species to fishing* and *selectivity and scale of operation of fishing patterns*. Finally in Chapter 6 we summarize and combine the findings of the previous chapters and try to identify the major implications for management. First we consider it from a biological perspective, before we turn to aspects more based on social considerations.

1.4 Methods and data

The report is based on ten case studies and a literature review. The case studies are to be considered as a part of the report, but they are published in a separate volume. The water bodies included in the case studies are: Lake Mweru-Luapula and Lake Bangweulu in Zambia; Lake Kariba covering both the Zambian and Zimbabwean sides; Lake Chilwa and Lake Malombe in Malawi. Ideally we would have liked to do combined fieldwork covering both biological and social science research on all the water bodies, but for practical and methodological reasons this was not possible. Each of the five case studies focusing on biological issues covers one of the five water bodies, while those focusing on social science issues only cover three lakes; three cover effort development issues in Kariba (Zambian side), Mweru-Luapula and Malombe, one covers the role of trade in Kariba (Zambian side), while the last case study is a comparative study of the history of fisheries regulations in Zambia and Zimbabwe.

The authors have considerable experience in management-related research related to small-scale fisheries development in the region. Kolding and van Zwieten have extended research experience from Kariba and Mweru, while Raakjær Nielsen has worked on South African fisheries. Overå and Jul-Larsen have most of their research experience from West African small-scale fisheries – marine as well as freshwater. In addition, all have short-term experience through missions, research and postgraduate teaching and supervision from a number of other lakes including Victoria, Tanganyika, Malawi, Cahora Bassa, Chilwa and Itezhi-tezhi, as well as the Bangweulu and Okavango swamps, just to mention water bodies in the SADC region.

In social science, historical data have been problematic to obtain. Except for Kariba, few studies have been done in the areas we cover and survey data are scarce and difficult to get hold of. Since the studies on Mweru and on fisheries regulations in Zambia and Zimbabwe are connected to projects for doctoral theses, archive studies have been undertaken in these, but not for the others. Effort data from the last two decades have mainly been collected from frame surveys where these existed and later adjusted according to other effort data compiled by the biologists (see below). The four case studies dealing with local conditions are based upon one or more pieces of fieldwork using various types of interview methods in combination with observation and discussions with assistants in the local areas. The study on trade followed traders between Lake Kariba and Lusaka. A small survey covering 460 fishing households on the Zambian side of Kariba was organized and made a valuable contribution to understanding the occupational mobility among the fishers. A follow-up of an inventory of fishing gear, first undertaken in 1993 and covering all the fishing units in Malombe was also organized.

Data used for the biological case studies are all from existing data collections stored in various formats at the field offices and statistical departments of the respective fisheries departments or research institutes. Biological and fishery data entail both time series of catch and effort data from Catch and Effort Data Recording Systems (CEDRS) and experimental survey data (Table 1.1). The CEDRS of the various lakes are either sampling based, i.e. two stage systems where catch rates are sampled in monthly (Kariba, Malombe, Chilwa), quarterly or tri-annual (Mweru, Bangweulu) surveys and total effort is counted in annual (Malombe, Chilwa) or irregular (Kariba, Mweru, Bangweulu) frame surveys. Other CEDRS's are based on logbooks (Kapenta fishery in Lake Kariba). Experimental data contain more biological detail but are mainly shorter or longer time series based on gillnet-surveys using experimental fleets of gillnets with a range of different mesh sizes (Kariba – Zimbabwe and Zambia – and Mweru). In both Bangweulu and Mweru, local fishermen were involved in experimental fishery data collections (see Ticheler *et al.*, 1998 for methodology). Fishermen recorded data daily for the pelagic light fishery on Chisense (*Microthrissa moeruensis*) in Lake Mweru. We refer to the individual case studies for a detailed discussion of the data and the various data collection procedures used in each lake. Lake-level data were obtained from the respective hydrological departments (Kariba, Mweru, Bangweulu, Malombe and Chilwa) supplemented by own measurements (Mweru) and literature (historical data).

Table 1.1 *Time series used in the biological case studies and data sources. CEDRS = Catch and Effort Data Recording System. In Kariba logbooks were used in the Kapenta fishery on both sides of the lakes. The logbooks of Lake Mweru refer to the sample of 21 Chisense fishermen who filled in daily logbooks over the period indicated.*

Lake	Lake levels	Experimental data	CEDRS Sampling	Logbooks	Data Sources
Kariba Zimbabwe	1962-1999	1970-1999	1973-1999	1974-1999	1,2
Kariba Zambia		1980-1999	1980-1999	1981-1999	3
Mweru	1955-1998	1970-1997	1953-1998	1994-1998	4,5,3,5
Malombe	1979-1998		1978-1997		6,7+8
Chilwa	1997-1998		1976-1997		6,7+9
Bangweulu	1956-1995		1952-1991		4,5

Zambezi River Authority, Lusaka, Zambia
Lake Kariba Fisheries Research Institute, Kariba, Zimbabwe
Department of Fisheries, Chilanga, Zambia
Department of Hydrology, Lusaka, Zambia
Department of Fisheries, Nchelenge, Zambia
Water Department, Lilongwe, Malawi
Monkey Bay Research Unit, Monkey Bay, Malawi
Mangochi Fisheries Office, Mangochi, Malawi
Kachulu Fisheries Office, Zomba, Malawi

A point of concern for many African small-scale fisheries, and equally for the cases in this study, is data reliability. A general tendency exists among researchers and development workers to dismiss catch and effort data as completely unreliable, possibly forged, and in most cases useless for anything but the roughest indications of catch and effort levels. This rejection is too easy. In all cases in this study it is possible to point out where and how data collections become unreliable, and in particular for what purposes they can and cannot be used (see also Chapter 5). Little evidence exists for data forging or false reporting in the cases studied; most errors, at least with the sampling systems, are "administratively induced", and as the Malombe case shows (Zwieten *et al.*, 2003a) they can be quantified based on the original sample data. We refer to each case study for a more detailed discussion of data reliability.

2. CONCEPTUAL CLARIFICATIONS

2.1 Fisheries concepts

2.1.1 Fishing effort, catch and catch rate

An understanding of fishing effort is fundamental for assessing and managing fish stocks. Most management principles involve deciding directly or indirectly upon the amount of fishing effort (f) that should be applied to the stock to obtain a certain amount of catch (C) that is sustainable over time (Rothchild, 1977). Furthermore, the most commonly used contemporary method of estimating the relative abundance of an exploited fish stock is by using the catch per unit effort (C/f) as an index of abundance.

The basic assumption in fisheries theory is that catch (*C*) and stock abundance, or standing biomass (*B*) are related by

$$C = q \cdot f \cdot B \tag{1}$$

where *f* is a measurement of the nominal fishing effort or intensity, and *q* is the so-called catchability coefficient (defined below). The nominal fishing effort is expressed in for example the number of fishermen, the number of boat-days, the number of meters of gillnet set, the number of hooks set, the number of pulls or shots made, etc. For fisheries data, however, it is generally difficult to measure the nominal effort precisely, and in particular to standardize it in terms of relative fishing power. Unfortunately, due to changes in the catchability coefficient *q*, there is no necessarily fundamental relation between the magnitude of the nominal effort and the magnitude of the catch. Therefore, for stock assessment purposes, there is a need for a measure of fishing effort that has a constant effect upon the fish population. This measure, commonly used in the population dynamics literature, is the so-called fishing mortality.

The fishing mortality (*F*) is simply defined as the fraction of the average population taken by fishing. In other words, *F* can be considered as an invariant measure of effort (Rothchild, 1977). *F* is also called the instantaneous rate of fishing mortality, i.e. the rate at which fish are dying due to fishing, and therefore expressed per time unit, usually per year. *F* can be measured without reference to the nominal effort, the configuration of the fishing gear, or the manner in which the gear is employed. *F* can be defined as

$$F = \frac{C}{B} = q \cdot f \tag{2}$$

Although *F* is defined as the fraction of the average population abundance taken by fishing, and therefore one would expect it to take values less than 1, it can in practice have a value of more than 1 on an annual basis for stocks with a high biological regeneration rate. This is because the annual *productivity* for such stocks, and therefore the cumulated annual catches, can be much higher than the average standing abundance (mean biomass). These are stocks – often smaller-sized tropical fish species – with a so-called high biological turnover, or high production to biomass ratio (*P/B*). These production concepts are further defined in Appendix 1.

The catch rate (*C/f*) or Catch per Unit of Effort (*CpUE*) is the catch per unit of effort over a time interval and defined as

$$CpUE = \frac{C}{f} = q \cdot B \tag{3}$$

For scientific research surveys, or experimental fishing, effort is standardized and fishing gears kept constant in order to keep a simple relationship between catch rates and population abundance (*B*), i.e. to minimize the inherent measurement errors and/or variations in *f* and *q*. However, this so-called fishery-independent monitoring of stocks through scientific surveys is expensive and surveys often cannot generate the amount of data needed for the evaluation of states or changes in fish stocks or mortality rates (fishing pattern), especially not in the highly diverse tropical freshwater systems discussed in this report.

In many parts of the world, the main supply of information on fishing effort, catch and catch rate is through monitoring of fisheries input (fishing effort) and output (catch), i.e. through fishery-dependent monitoring. Long-term monitoring of fish stocks therefore is almost by necessity dependent on information obtained through the fisheries exploiting them and with that on the official fisheries statistical system in use. Fishery-dependent monitoring entails at least the collection of two essential parameters in fisheries statistics: catch (*C*) and fishing effort (*f*) and from these the derivation of catch rate (C/f) (FAO, 1999). CEDRS maintained to address information needs for fishery management vary in their degree of administrative and statistical sophistication, but all share the collection and maintenance of these basic parameters.

2.1.2 Catchability

Catchability (q) is defined (see equation 3) as the relationship between the catch rate ($CpUE$) and the true population size (B). So the unit of catchability is *fish caught per fish available per effort unit and per time unit*. Catchability is also called gear efficiency (Hilborn and Walters, 1992) or sometimes fishing power, and is strongly related to gear selectivity (defined below) because it is species and size dependent. Sometimes gear selection is simply defined as the relative change in q (Godø, 1990). Therefore the fishing mortality (F) as a function of the size (length) of fish, i.e. the fraction of fish caught per fish in the population, has the same shape as the gear selection curve (S), but with a different value depending on the nominal effort.

In other words, when effort (f) is equal to 1 (unit) then:

$$q = F = S \qquad (4)$$

This means that q can conceptually be considered as the probability of any single fish being caught. Therefore q ranges between 0 and 1.
However, the probability of a fish being caught at any time depends on several factors, which are not only man-made, and can broadly be grouped as biological or technological:

Biological factors include:

- fish availability on the fishing ground
- fish behaviour towards the fishing gear
- the size, shape, and external features of the fish
- where some of these factors again are depending on season, age, environment and other species.

Technological factors include:

- gear type, design, size, colour and material
- gear position, duration and handling
- experience of the fishermen
- where again these factors are depending on biological changes.

As both the unit and the different notation show, the catchability coefficient (alias efficiency, or fishing power, or probability of a fish being caught), is therefore a composite and very complicated factor. Conceptually, however, 'fish catchability' implies primarily changes in fish behaviour (May, 1984), whereas 'fishing efficiency' indicates changes in fishing practices (Neis *et al.*, 1999) or in relative fishing power. As information on the possible causes of variation in q is normally lacking, the biological and technological factors are for practical purposes normally assumed invariant of abundance, time, species, size/age, and the individual skills of the fishermen. However, only under this very rigid assumption can the catch rate ($CpUE$) be considered directly proportional to the stock abundance (Hilborn and Walters, 1992) and be used as an index of the stock size.

Consequently, the catchability (q) cannot be quantified directly if catch rates are used to estimate stock sizes. The standard solution to evaluate changes in efficiency (fishing power) in a fishery over time, and with that the catchability, is therefore to compare catch rates from commercial and research fishing where the catchability of the research fishing is holding constant from year to year (Neis *et al.*, 1999):

$$\frac{CPUE_{fishery}}{CPUE_{research}} = \frac{q_{fishery}}{q_{research}} \qquad (5)$$

This method requires several years of data in order to detect relative changes in the efficiency of the commercial fishery. This lag in time, before eventual changes are discovered, will lead to overestimation of stock size if the commercial fishing efficiency or fishing power is rising (Pope, 1977).

The variability and elusiveness of the parameter q, and the difficulties in quantifying it, is a very important reason for the difficulties in analysing the relationship between the magnitude of nominal effort (which is also difficult to measure) and the direct effect on the regenerative capacity of the stocks. Changes in q, which are mostly unaccounted for, induce additional uncertainty in the parameter catch rate as an index of stock abundance, if q simply varies through time. If there is a unidirectional change in q, as a result of, for instance, increased fishing power of a unit of effort over time, catch rates – and with these stock abundance - will be systematically overestimated. This is also an important reason why, to a large extent, fisheries science is operating with the parameter 'fishing mortality (F)' instead of the parameter 'fishing effort (f)'. Unfortunately, fishing mortality is notorious for its incomprehensibility outside the fisheries scientific community. However, returning to equation (2), the elusiveness of F, and its relationship with nominal effort and efficiency in a biological sense (i.e. the efficiency with which a fish is caught) may become somewhat less obscure when illustrated graphically (Figure 2.1).

2.1.3 Selectivity

A generally important technical measure for fishing gears is the size selectivity which is defined as the probability of fish being retained in a fishing gear as a function of the length of the fish (Misund *et al.*, 2001). These probabilities are often expressed as various mathematical models. A selection curve (i.e. the probability of capture plotted against the size of the fish) for trawl gears is mostly sigmoid or S-shaped, whereas bell-shaped curves are normally the case for gillnets and hooking gears. Important selectivity measures are L_{50} – defined as the length of the fish where the fish have a 50% probability of being retained by the gear on encounter – and the *selection factor* – defined as L_{50} divided by mesh size in cm. In addition to the *selection range* which is defined as $L_{75} - L_{25}$ (L_{75} is the length of the fish where 75 % of the fish is retained, and L_{25} is the length where 25 % of the fish is retained), these parameters describe the size selection characteristics of fishing gears.

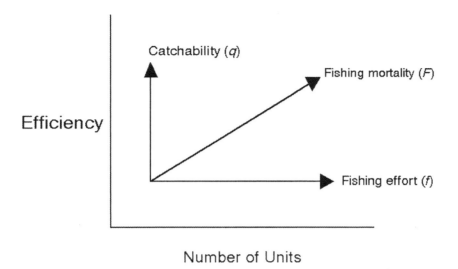

Figure 2.1 *Fishing mortality (F) as a resultant of nominal fishing effort (f) and catchability (q). The number of units expresses nominal fishing effort, while catchability can be expressed by the efficiency of one unit of fishing effort*

All fishing gears are species and size selective: this is also true in multispecies fisheries where one type of gear may catch a set of species, while another gear or the same gear used in a different way or different area may catch another set. This means that all fishing gears are only able to catch a certain portion of the total (multispecies) fish community present. The use of the catch rate as an index of abundance of a fish stock is therefore further complicated by the selectivity of a fishing gear. Catch rates only reflect the abundance of the *fishable stock* or, in other words, that portion of a fish population or fish community that can be caught by a specific gear. Catch rates can be used as an index of abundance for the total stock, under the assumption that all specimens within a (multispecies) stock at some stage during their life become part of the fishable stock.

The area of operation of a gear, the inconstant behaviour of the fish relative to the gear, and the size of the fish determine the part of a stock that can be caught by a gear. As discussed, these factors are all included in the parameter catchability. Selection may therefore differ in different areas of operation simply because of the species and sizes present on a fishing ground. For example, seines with small mesh sizes operated in shallow areas near shore that act as nursery grounds will have a higher probability of catching juvenile specimens than if they are operated further offshore where these fish are not present. Species behaviour may change seasonally as a function of several factors, such as migrations, spawning or temperature, and with this the probability of catching it with a certain gear will also change. Furthermore, species may develop avoidance behaviour towards gear, which will result in a lower catchability. Some species are notorious for their ability to avoid certain gear, for example the ubiquitous redbreast Tilapia is very difficult to catch with gillnets.

Fishing gears are intrinsically associated with selectivity, and selectivity, or the impact of fishing on an ecosystem, is an essential component of a management programme. The importance of selectivity is therefore a key point for most researchers and managers, and any non-selective capture method automatically carries the connotation of being harmful or destructive, or at least leading to growth-overfishing seen from the traditional single-species perspective. Mesh size and gear restrictions are therefore among the most easily applied and widely used management regulations. Consequently most nations have imposed legislation that bans certain gears and mesh-sizes with the aim of protecting the resource (Gulland, 1982). Although many of these regulations originate from problems associated with the large scale-fisheries (Misund *et al.*, 2001), they are often uniformly applied in all sectors. However, selectivity seems much more of a problem for industrialized fisheries that dump on average about 45% of their catch, while small-scale artisanal discards average only 5% (Bernacsek, 1989) despite the fact that they mostly operate in more multispecies environments. Although numerous authors have already pointed to the problems of defining the "right" mesh-size in a multispecies fishery, the notion of regulations on selectivity still persists. In addition, small-scale fisheries, such as those studied in this report, often use a variety of gears, both traditional and modern. Many of these gears and particularly the traditional ones, such as seines, small mesh-sizes, drive- or beat fishing, barriers and weirs, are often classified as illegal under the pretext of being non-selective with assumed negative impacts on the fish populations.

2.2 Social and economic concepts

2.2.1 Population- and investment-driven changes in fishing effort and demographic change

The preceding section is based on fisheries theory and definitions. It is open to question, however, whether the concepts of 'nominal fishing effort (f)' and 'catchability (q)' – in particular where it deals with technological factors and their biological impact on the fish stocks – can be readily applied within social and economic treatments of fishing effort. In fisheries economics, for example, fishing is analysed in terms of investments and revenue, and 'effort' is therefore conceptually associated with cost. According to Hannesson (1993), it is generally impossible to develop any general valid proportion between the nominal effort in terms of units and fishing effort in the economic sense due to technological differences and/or changes and different inputs of fuel, equipment, manpower, etc. Therefore in fisheries economics, fishing effort (investment) is sometimes defined as the mortality generated on a fish stock because that will more accurately predict the catch (or revenue). This makes effort strictly equivalent with the fishing mortality (F) used in population dynamics:

$$C = F \cdot B = ? \cdot f \cdot B \qquad (6)$$

Comparing with equation (2), fisheries economics then appear to have similar problems with the relationship between the nominal effort and the catch as in fisheries biology with the elusive catchability coefficient q.

As for fisheries economics, Brox (1990) does not distinguish between nominal effort and fishing mortality. However, as a social scientist, Brox is concerned with identifying how different ways of changing effort may lead to very different social results. He therefore introduces the distinction between *'horizontal'* and *'vertical'* changes in fishing effort (or fishing mortality). Horizontal growth is related to the growth in the number of fishermen: "Fishermen's children grow up and establish households that base their economy on participating in fisheries.... Fishing communities also absorb people born outside fishing districts, especially through marriages, and young migrants initially looking for temporary employment"(Brox, 1990:233). It is important to note that horizontal growth is different from demographic growth. While demographic growth in the traditional meaning of the concept is understood as a function of birth- and mortality rates, horizontal growth of effort also includes migration and changes in people's occupations.

Vertical growth is related to growth in capital use and technological level. "The aggregate effort may also increase because each operating unit acquires more technical equipment. This happens when some of the formerly stagnant units begin to accumulate capital, but also as investors from other branches of the economy, other regions or even other countries somehow decide to participate in harvesting the resources." (Ibid.)

These concepts, which for the sake of clarity we have renamed *population-driven* and *investment-driven* growth in fishing effort, seem to come very close to the above distinction between nominal changes of effort and changes in efficiency. From a biological point of view, however, it is the interaction between the catching process (gear) and the fish that is important. Thus, with regard to 'numerical growth of fishing effort', it is the total number of units of gear and their employment that matters. Who invests and at what level is irrelevant for the fishing mortality: 10 fishermen owning 100 nets each is the same as 100 fishermen owning 10 nets each. Investment-driven change in effort is therefore not directly equal to the biological term 'catchability', but more related to 'efficiency'. The two concepts will only coincide if the unit of fishing effort is understood to be production units and if the biological variables (e.g. behaviour) included in catchability are excluded.

The importance of distinguishing between socio-economic processes leading to simply numerical changes in nominal fishing effort or increased efficiency from capital investments, lies in the different effect each of these processes has on the sustainability of exploitation. The hypothesis is that when investment-driven growth of effort accelerates, it represents a potentially much bigger growth in absolute fishing mortality than that expected to be caused by demographic processes. This expectation is based on the presumption that investment-driven growth lead to technological changes towards more effective gear, whereas population-driven growth, just leads to 'more of the same'. It is important not to consider population-driven changes in effort as a mere reflection of the *demographic changes* in a community. As mentioned above, demographic change is generally defined as the difference between birth rates and death rates in a given population, but population-driven change in fishing effort – as indicated by Brox – also includes changes due to migration or to changes in occupations. As will be shown, these variables are sometimes more important than the demographic changes.

2.2.2 Social institutions

Institutions constitute a central element in analysis of effort development. In fisheries management literature, institutions are often seen as government's way of organizing their fisheries administration. We see institutions as much more than this. In recent years the 'neo-institutionalist' approaches in economy and political science (e.g. Elinor Ostrom, Douglas North, Oliver Williamson and Jean-Philippe Platteau) has dominated the understanding of social institutions. A commonly used definition is: "*Institutions are rights and rules that provide a set of incentives and disincentives for individuals geared to minimize transaction costs.*"

Several scholars have pointed out that the focus on transaction costs and market imperfection is too narrow and neglects the historical dimension in how rules have been constructed. From a sociological perspective, Scott (1995) brings in the cognitive aspects and argues that institutions are cultural, social structures and routines. From social anthropology, Berry tends to view institutions "...not as the rules themselves, but as regularised patterns of behaviour that emerge from underlying structures or sets of 'rules in use'. ... Rather than existing as a fixed framework, 'rules' are constantly made and remade through people's practices" (Leach et al., 1999: 237). The works of, for instance, Berry (1989, 1993), Bierschenk and Olivier de Sardan (1998) and Peters (2000), convincingly argue that local institutions crucial in the regulation of people's access to vital resources are often unclear and even lack coherence. Their research shows that so-called 'local management systems' have often emerged as a result of negotiation and accommodation in long-term power struggles. Various actors, including a variety of external powers operating at different periods of time, have based their claims and legitimacy on different 'logics' and values. The lack of hegemonic power at the local level has facilitated their continued co-existence and has entailed an institutional landscape where rules are characterized by ambiguity and sometimes even by contradiction.

Our task here is not to support either the neo-institutional or the more constructivist approach to institutional analysis, but simply to point out the differences and alternatives that they represent in the analysis of regulation and access to natural resources. In the same way that the 'management' and the co-management approaches represent alternative ways for understanding existing management regimes, studies of African fisheries have shown how both approaches have important contributions to provide (Jul-Larsen, 1999). While the neo-institutional approach may shed light on the fact that artisanal fisheries sometimes seem to out-compete more industrialized fisheries, the constructivist approach may help us understand important relationships between the local institutional landscape and prospects of economic development.

We will also emphasize that institutions can be either formal or informal and may be created or evolve over time. The nature of institutions can be explained by the use of an iceberg analogy. The top, visible part of the iceberg can be taken as the formal and written institutions, whereas the lower part of the iceberg, which is not visible but exists, can be taken as the informal and unwritten institutions. Both formal and informal codes may be violated and therefore punishments are enacted. The essential part of the functioning of institutions is thus determined by whether their codes can be enforced, the cost of enforcement and the severity of the punishment.

3. TRENDS IN YIELDS AND FISHING EFFORT OVER THE LAST 50 YEARS

In order to put present effort developments in SADC freshwater fisheries in a historical context, this chapter will seek to identify some of the trends in the harvesting, processing and trade of fish from the lakes we have studied. Who are the people who become fishers, how many are they and why do they choose fishing as an occupation? How and to what extent did the fisheries develop into production for commercial exchange and what have the consequences been? What is the relation between the fisheries and other economic sectors? How stable are the communities, markets and macro-economic and political circumstances that condition developments in these fisheries? Before we look into the causes behind changes in effort in more detail in the next chapter, these are some of the general questions we will address in order to get a better picture of how fishing effort has developed during the second half of the 20th century.

3.1 Trends in yields and effort in the SADC region

3.1.1 Trends in yields

There are many practical and methodological problems involved in obtaining reliable catch and effort data for the SADC area at an aggregate level. According to FAOSTAT (FAO, 2000a and b), total yields in the freshwater fisheries of 12 mainland SADC countries increased from 168 000 MT in 1961 to 635 000 MT in 1997 (Figure 3.1). However, the increase was not linear but mainly took place in two periods (from 1961 to 1970 and from 1980 to 1990) with more stable periods in between.

If we look at the yields of individual countries in the same time period (1961-1997), it appears that out of the 12 countries, there are four s that account for more than 90 % of the total annual yields in the region: the Democratic Republic of the Congo, Malawi, the United Republic of Tanzania and Zambia.

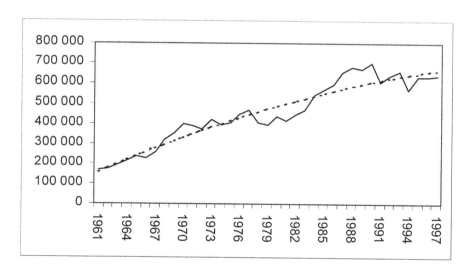

Figure 3.1 *Total annual yields in the freshwater fisheries of 12 mainland SADC states 1961-97 (in tonnes). The dotted line is the two-order polynomial trend line indicating a slight decline in the overall rate of increase.*
Source: FAO, 2000a.

Table 3.1 *Annual freshwater yields (average over five-year periods) by country in 12 mainland SADC countries (in tonnes).*

Country	1961-65	1966-70	1971-75	1976-80	1981-85	1986-90	1991-95
Angola	5 520	5 960	6 400	7 100	7 500	8 000	6 800
Botswana	480	720	1 180	1 270	1 420	1 860	1 960
Dem. Congo	66 600	104 720	116 506	107 640	119 160	159 700	169 835
Lesotho	0	0	7	25	18	28	36
Malawi	22 120	35 980	68 114	67 331	60 786	77 022	62 819
Mozambique	1 600	2 100	2 800	3 200	3 500	3 475	4 501
Namibia	0	0	20	50	260	980	1 114
South Africa	100	310	1 150	1 150	1 017	1 759	2 263
Swaziland	0	0	0	45	82	106	118
Tanzania	66 500	116 340	146 975	182 955	218 686	319 565	281 529
Zambia	33 020	43 200	50 291	51 213	58 843	64 787	67 722
Zimbabwe	1 560	2 000	3 048	8 051	16 271	22 025	20 424
Total	197 500	311 330	396 491	430 030	487 542	659 308	619 121

Source: FAO, 2000a

No aggregate statistics on the species composition of the catches exist. However, it is beyond any doubt that the catch composition has changed towards an increased share of small and fast-maturing species. At present, the catching of small pelagics or other small-sized fast-maturing species at low trophic levels (i.e. low in the food chain) is becoming an increasingly important part in the fisheries of all the major lakes in the SADC. One can mention Dagaa in Lake Victoria, Kapenta in Lake Tanganyika and Lake Kariba and Chisense in Lake Mweru. These species have not always been as important as they are today in terms of employment generation and nutrition and 40 years ago few of these stocks were systematically exploited. This trend is a major feature of SADC freshwater fisheries at present (see Chapter 5).

3.1.2 Trends in effort development

If good yield data are difficult to obtain, reliable statistics on fishing effort are even harder to get hold of. However, some partial data sets exist that can give us some indications. George Coulter has for several years assembled long time series of the increase of fishers and fishing boats in African freshwater bodies, which are based upon data from a number of various primary and secondary sources. Supplemented by some of our own data on 14 lakes in Central and Southern Africa, his findings are presented in Table 3.2.

These data tell us at least three interesting things. First they show how fishing effort has grown: the number of fishermen has increased by approximately 160% over a period of 20 years and the number of boats has increased by about 70%. Secondly, it shows that the growth in effort seems stronger in terms of increases in people than in fishing gear (boats). Thirdly, but not least important, they demonstrate the large variations in effort development, even for water bodies situated close to each other. While Lake Mweru saw the number of fishermen increasing by 215 percent over 20 years, Bangweulu, located close by (see map), experienced a reduction of fishers in the order of 24 percent.

Table 3.2 *Development in numbers of fishermen and boats from the early 1970s to 1989-92 in 14 African lakes. Yields, density and catch rates refer to the 1989-92 period. Names in bold indicate the systems included in this study.*

Lake	Area	Yields late1980s	Effort early 1970s		Effort 1989-92		Percent Change		Density	Catch
	Km^2	t/year	Fishers	Boats	Fishers	Boats	Fishers	Boats	Fishers/km^2	t/km^2
Victoria	68 800	187 495	26 000	9 643	105 000	21 986	304	128	1.53	2.73
Tanganyika	32 600	73 000	15 000	9 500	40 000	10 887	167	15	1.23	2.24
Malawi	30 800	28 000	10 154	4 000	27 296	10 260	169	157	0.89	0.91
Bangweulu	7 900	10 900	13 400	5 475	10 240	5 900	-24	8	1.30	1.38
Turkana	7 570	1 500	1 850	106	1 500	141	-19	33	0.20	0.20
Kariba	5 364	30 311	1 600		7 060	2 244	341		1.32	5.65
Albert	5 270	28 230	9 600	2 626	20 442	3 437	113	31	3.88	5.36
Mweru*	2 700	42 000	6 000	4 155	15 791	6 600	163	59	5.85	15.56
Kivu	2 699	315	600	500	2 868	981	378	96	1.06	0.12
Edward	2 300	16 031	5 700	400	5 443	1 134	-5	184	2.37	6.97
Chilwa	750	15 000	1 740	700	3 485	2 030	100	190	4.65	20.00
Malombe	390	7 500	900	500	2 371		163		6.08	19.23
Itezhi-tezhi	370	1 200	290	253	1 250		331		3.38	3.24
Chiuta	113	1 400	193		350		81		3.10	12.39
Total/mean	167 626	442 882	93 027	37 858	243 096	65 600	161	73	1.45	2.64

Sources: Bayley, 1988; Bossche and Bernascek, 1990; Coulter, unpublished data; Cowx and Kapassa, 1995; Greboval *et al.*, 1994; Kolding, 1989; Kolding *et al.*, 1996; Mbewe, 2000; van Zwieten *et al.*, 1995.
* Only covers the Zambian part of the lake.

One should be aware though of the many pitfalls in the interpretation of aggregate effort data. Firstly, there is the problem of comparing data that have been collected for different purposes and with different methods. Secondly, increases in the number of fishermen and boats cannot be uncritically interpreted as increases in fishing effort on the same resources. Thirdly, at the country level, data generally does not show when new or previously unused water reservoirs are being included (e.g. man-made lakes such as Lake Kariba and Cabora Bassa). Finally, aggregate effort data do not show if fishing in the same water bodies has expanded and started targeting a broader range of species[2] (see Chapter 5). As mentioned above, most fisheries now include a range of previously untouched stocks, but even when taking these modifying factors into account, there can be little doubt that fishing effort – also on the already exploited stocks – has increased substantially in SADC freshwaters. As will be shown below, our own research on the development in specific lakes and water bodies also supports this view.

However, aggregate data may also sometimes reveal correlations, which no one is aware of and which may prove significant for understanding effort dynamics. The data from Table 3.2 show a close relationship between the density of fishers and the average catch rates in the different water bodies (Figure 3.2). Although effort and catch are not independent figures (see Chapter 2), the relationship, taken across different water bodies with few examples of declining yields, may indicate that the observed increase in effort on the whole appears sustainable. In addition Figure

---

[2] For example, the 1989-92 figures for Lake Kariba in Table 3.2 include 1 232 Kapenta fishermen. Kapenta (*Limnothrissa miodon*) was stocked into Lake Kariba in 1967 and the fishery was opened in 1974.

3.2 also reveals a remarkably constant average yield of about 3 MT per fisher per year, irrespective of water body. We can think of several hypothetical reasons for why such constant average yields emerge, but a number of practical reasons made it impossible to include the issue in our research. It should also be noted that three of the case studies in this report (Malombe, Chilwa and Mweru) have the highest observed relative effort and yields within the 14 different lakes.

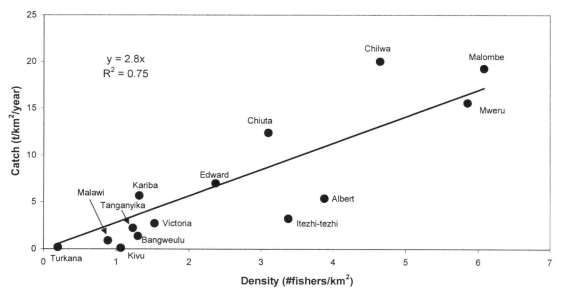

Figure 3.2 *Catch rates (annual catch/km^2) plotted versus effort density (#fishers/km^2) in 14 African lakes. The trend line indicates a relatively constant average yield of about 3 MT per fisher per year irrespective of water body and country.* Data and sources: see Table 3.2.

In order to get a better understanding of how trends in particular lakes fit with the overall picture of effort development, we have assembled and examined effort data in the water bodies selected in this study[3]. This examination supports the great spatial variations shown in Table 3.2. In Lake Mweru, effort has steadily increased since the 1950s and seems to continue to do so (Zwieten *et al.*, 2003b), while in the Bangweulu Swamps effort has probably remained fairly stable over a very long time (Kolding *et al.*, 2003b). A more complex pattern also emerges showing that fishing effort may vary considerably over time. In the Zambian part of Lake Kariba, fishing effort has varied considerably since the creation of the lake in the late 1950s. Until 1963, fishing effort increased very fast, but then it fell sharply for some years before it slowly began to grow again. In the 1970s, the fishing effort dropped again only to start increasing considerably in the 1980s. In the 1990s, effort started falling again, and on the Zambian side today it is probably not much higher than it was just after the lake was filled.

If we take a look at Lake Malombe, fishing effort increased steadily from the 1950s onwards, but around 1990 it stabilized and in recent years it has decreased quite drastically (Zwieten *et al.*, 2003a). Lakes such as Chilwa (Zwieten and Njaya, 2003), Chiuta and Mweru Wa Ntipa are also

---

[3] This has mainly included Lakes Mweru (the Zambian side), Kariba (the Zambian side), Malombe and Chilwa, but it also includes closer examinations of secondary data from Lakes Tanganyika, Chiuta, Mweru wa Ntipa and the Bangweulu swamp.

subject to considerable fluctuations in effort. In order to provide a better understanding of these past and present trends, we will describe the historical development of effort in three of the lakes: Lake Mweru, Lake Kariba and Lake Malombe.

3.2 Three lakes: different histories – different effort trajectories

3.2.1 Lake Mweru

Until the development of copper mines in Northern Rhodesia (today's Zambia) and Belgian Congo (today's Democratic Republic of the Congo) in the 1920s and 1930s, fishing in Lake Mweru and in the Luapula River was an integrated part of the rural economy in the Luapula Valley. The British and Belgian colonial powers had quite different interests and procedures in their development of the fisheries on the two lakeshores. However, in both countries local African leaders were given some kind of authority to control access to the resources of the fisheries. Fishing effort at that time largely fluctuated according to the number of people engaged in fishing and these fluctuations reflected the periodical needs of the local population to supplement the production of staple products, such as millet and cassava. As David Gordon puts it in his case study: "The fishery was a lifesaver in times of trouble. When warfare interrupted stable farming, the river and the lake provided some sustenance" (Gordon, 2003).

When the copper mining industry developed in both countries, the fisheries and fish as a commodity were to take on a new role. It became crucial for the industrial entrepreneurs, and for the colonial authorities to assure that fish could provide cheap, protein-rich food for the labour force in the mines. The companies built roads to connect Lake Mweru to the mining cities and began to contract traders, mainly of foreign but also of local origin, to supply their workers with fish. This increased demand for fish resulted in a considerable growth in numbers of fishermen on both sides of the lake. The bulk of these fishermen seem to have been people already living in the area. Even if many of the richer traders invested in ice plants and means of transport, they did not initially participate in the harvesting of fish. On the Northern Rhodesian side, the British maintained that the fishery should be reserved for the Africans, while the Belgians in Congo initially had less clear policies. However, when World War II broke out and food insecurity increased, the Belgians started to promote European investments in production in order to get as much fish out of the lake as possible. The Belgian mining company also supported hunting campaigns to eradicate crocodiles that had made large-scale investments in fishing equipment difficult. On the Congolese side, expatriates (mainly Greeks) started investing in larger vessels and in large bottom set-nets in the spawning areas of the Mpumbu (*Labeo altivelis*). In the years just after the war, the external investments were considerable and resulted in a collapse of the Mpumbu fishery within few years. In 1950 the fishery and the industrial fleet thus came to an end. But the over-exploitation of Mpumbu did not affect the fisheries of other species very much. Moreover, eradication (or at least a substantial reduction) of crocodiles supported the local fishery, which during this period became more and more concentrated on the catching of Pale (*Oreochromis mweruensis*) and other ciclids with gillnets. The demand for fish in the mining district remained high and the fishery continued to grow through a steady increase of new fishermen throughout the 1950s and early 1960s.

As a result of Zambia's independence in the 1960s, the copper industry was nationalized. Shortly after, in the early 1970s, copper prices on the world market fell drastically, and this led to considerable recessions in the copper mining towns and markets. The fisheries of Lake Mweru were affected in various ways. On one hand, investors with capital became less interested in the fisheries: ice plants were closed down, and the marketing of fish was increasingly left in the hands of small-scale traders of dried fish. The fishery thus became a "lifesaver" again, and we must assume that the reduced levels of investment led to a certain "technological recession".

The reduced opportunities in the industrial sector also had another very different effect. As a result of the crisis in the copper industry, numerous laid-off mine workers, as well as many other people who had earned a living in the urban service economy, returned to the rural areas seeking access to resources where they could find them (Ferguson, 1999). Many of these men and women came from the Copperbelt to Lake Mweru to look for opportunities in the fishery. Since the catch rates in the gillnet fishery had shown very clear decreasing trends throughout the 1970s, many of the new entrants looked for opportunities in a different niche – the Chisense (*Microthrissa moeruensis*) fishery and trade which emerged as a more commercial enterprise in the early 1980s. The new entrants, some of whom had been able to accumulate funds in their previous jobs, apparently found it easier to experiment in a fishery targeting new stocks and utilizing new technologies than to enter into direct competition with the existing fishermen. Many women found a new opportunity in trade of the small Chisense, which was a product that suited the increasingly poor urban consumer market very well. Some of these traders were also able to invest in fishing gear and employ men (often their sons) to fish Chisense for them (Gordon, 2003). Although the Chisense fishery never came to dominate on Lake Mweru, it came to play an important role in the 1980s and 1990s and facilitated the continued growth of effort in general and numbers of fishermen in particular. The gillnet fishery for Pale and other species remained the dominant catch method and continued to grow substantially both in terms of fishermen and nets (Figure 3.3).

In the 1990s, the Zambian government implemented harsh programmes of structural adjustment, but the urban economy in the Copperbelt continued to deteriorate. The fisheries of Lake Mweru thus continued to provide an option for a continuing flow of new entrants in need of a livelihood. Frame survey data from this period indicate a very high turnover of people. While the number of fishermen between 1992 and 1997 is reported to have increased by about 2 300 individuals, about 5 400 people interviewed report that they started fishing in this period. This means that 3 100 fishermen must have left the fishery during the same period (Zwieten *et al.*, 2003b). Despite the steady and long-lasting increase in the gillnet fishery, the 1990s experienced an increase in the catch rates of tilapia at the same time as the Chisense fishery thrived. The larger specimen of tilapia had been only moderately targeted for many years due to a continuous reduction in the mesh sizes of the gillnets (Ibid.).

Some renewed interest by processing and distribution companies in investing in ice plants and marketing facilities can be noticed in the 1990s. Currently the investors are largely Zambian (though the largest investor is the son of the most powerful entrepreneur in the fishery on the Congolese side). These companies' main markets are the Copperbelt and Lusaka, as well as the much more lucrative but risky markets in the Democratic Republic of the Congo where fish can be exchanged for diamonds (Gordon, 2003). Some degree of increased investments has resulted in credit-supply contracts between processing companies and fishermen. However, it is worth noticing that the increased growth in production in the 1990s has mainly been a result of increased numbers of fishermen joining the fishery, rather than a more efficient technological level of production, as was the case in the 1950s.

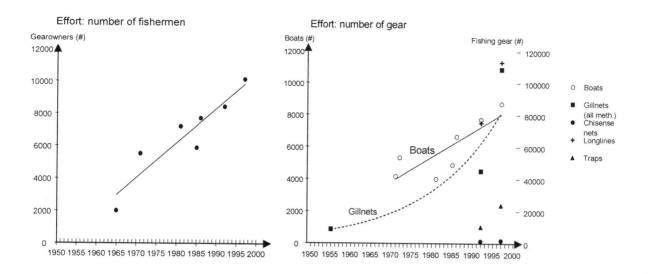

Figure 3.3 *Effort development: Fishermen, boats and gear in Lake Mweru.*
Source: Zwieten *et al.*, 2003b.

3.2.2 Lake Kariba (Zambia)

Since Lake Kariba is a man-made lake completed in 1958, the history of its fisheries is a recent one. Unlike other swamp and river systems in Zambia, e.g. the Kafue plains or the Bangweulu swamps where commercial fisheries developed before independence, fisheries in the Zambezi River remained marginal and at a subsistence level until the creation of the lake. In the late 1950s the colonial authorities, including the Gwembe Valley Native Authority, prepared for a major fishery initiative in the lake: fishermen were trained, sales points established and equipment and fishing gear imported. A European company attempted to get access to the fishery, and they went as far as to invest in an ice plant. But due to the African nationalist political climate during the years when Zambia was on the verge of achieving independence, the Europeans did not get fishing concessions. Instead, a simple gillnet fishery was reserved for the local Tonga population.

As a result of the high productivity during the filling phase of the lake, many people entered the new fishery. In the early 1960s the number of fishermen had already reached about 2 500. However, after the lake had reached its maximum level in 1963 and catch rates started to decline, the number of fishermen fell almost as quickly as it had increased in the years before (Figure 3.4). Most of the young Tonga men who had opted for fishing five or six years earlier now decided to return to the type of agro-pastoralism they had practised before, or to various forms of work migration. It was not until the late 1960s that the number of fishermen started to grow again, and this time at a more modest speed (see for example, Colson, 1971 and Scudder, 1965).

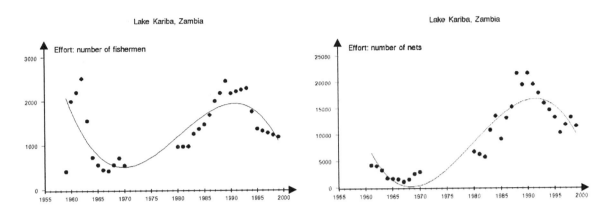

Figure 3.4 *Effort development: Fishermen and nets in the Zambian inshore fisheries Lake Kariba.* Source: Kolding *et al.*, 2003a.

Political independence in 1964 had led to the abolition of the Native Authority institutions, and this meant the end of the exclusive fishing rights that the Tonga had acquired in the 1950s. Slowly, "foreigners" – particularly from the northern areas of Zambia – started to dominate the fisheries on the lake. This process went on undisturbed until 1974 when the Zimbabwean war of independence made it very difficult to live and work along the northern shores of Lake Kariba. The fishery was formally closed by the Zambian authorities in 1975 and was only reopened in 1980 after the termination of the war the preceding year. Poor infrastructure, which deteriorated dramatically during the Zimbabwean liberation war, and a similar macro-economic context as described above for Lake Mweru, made marketing of fish from Lake Kariba cumbersome. Fish traders did not make enough profit to accumulate capital to invest in fisheries, and the area – due to its isolation – was unattractive for investors.

During the 1980s, the number of fishermen in Lake Kariba increased dramatically. By 1989 it was almost back to 2 500 (where it had been in 1963) while the number of nets had become almost four times higher compared with the initial period (see Figure 3.4). Except for the increased number of nets and a tendency towards smaller meshes (Scholtz *et al.* 1997), the technology of the inshore fisheries did not change much. And as before the war, it continued to be mainly non-Tongas from other parts of Zambia who fished the lake. Surveys undertaken in 1987 show that 60-70% of the fishermen were of ethnic origins other than Tonga (Walters, 1988). Although the technology remained very much the same, the new drive of the 1980s also led to a substantially higher number of nets per fishermen, which has since this time fluctuated between six and ten nets (Jul-Larsen, 2002).

However, 1980 was also the year when the Kapenta fishery started. Kapenta from Lake Tanganyika had initially been stocked into the lake in 1967 and 1969. It rapidly spread all over the lake and in 1974 a fishery was opened on the Zimbabwean side, but for security reasons it was not possible to open it in Zambia until the war of liberation ended. The Kapenta fishery was established as a much more capital-intensive production system compared to the inshore gillnet production. It requires rigs, deep-going dip-nets, electric light and winches and thus demands capital investments and knowledge about modern institutions and technologies far beyond the means of the average inshore fisherman. As in Zimbabwe, investments in the Kapenta fishery mainly came from white (Zambian, Zimbabwean or South African) entrepreneurs. The first Kapenta companies started operating in 1981, and catch rates were high. Kapenta is a small pelagic species, which is highly suitable as a commodity in poor segments of the urban markets. In order to reach these markets, the Kapenta operators relied on traders (mainly from Lusaka and the Copperbelt) for the distribution of bags of dried Kapenta (Overå, 2003). Kapenta trade thus became another welcome employment opportunity under deteriorating macro-economic circumstances.

In the 1990s, effort remained relatively stable in the Kapenta fisheries. It could have been expected that the operators, who generally have far better access to capital than their counterparts in the inshore fisheries, would try to increase their profits through technology development and increased investments. Many of them have access to credit through their enterprises in other sectors of the economy, through access to bank loans or through loans from business partners abroad. Interestingly, however, whereas the Kapenta operators on the Zimbabwean side have continuously improved their harvesting techniques (larger rigs, mechanization of the hauling of nets, hydro-acoustic equipment, etc.), the operators on the Zambian side of Lake Kariba have largely continued to operate on the same technological level as when they started 20 years ago. In order to balance their costs, they tend to keep their capital investments as low as possible. For example, they claim that the cost of labour (including theft of Kapenta by traders in collaboration with the workers) is so high, and the market price of Kapenta so fluctuating and limited, that increased capital investments are not really profitable.

In the inshore gillnet fishery, fishing effort decreased in the latter half of the 1990s. After stabilizing in the early 1990s, the number of fishermen gradually declined, and in 1999 it was reported that the number was only marginally higher than when the fishery reopened in 1980. A situation similar to that experienced in Lake Mweru, with a high mobility of people in and out of the sector, thus seems to be the case in the fisheries of Lake Kariba as well. While the total number of fishermen had dropped from approximately 2 200 in 1990 to 1 200 in 1998, a survey undertaken in 1998 reports that as many as 500 fishermen started to fish during the same period. In other words, at the same time as "new" people became fishermen, 1 500 fishermen left the Kariba inshore fishery during the 1990s. At the same time, an increasing number of local Tonga returned to fishing. Hence, according to a frame survey from 1995, the share of non-locals in the fishery had been reduced to about 55% of the total (Chitemburre, 1996). In addition, there is good reason to believe that the Tonga were more involved in the fisheries in the late 1990s than the 1995 frame survey suggests (Jul-Larsen, 2002). Therefore, at the same time as the number of "foreign" fishermen has decreased, for the local valley population the importance of the inshore fishery, often in combination with agriculture, has increased.

3.2.3 Lake Malombe and the South East Arm of Lake Malawi[4]

In the South East arm of Lake Malawi and in Lake Malombe, fishing used to be an occupation that was combined with agriculture. In 1915 it was reported that fishing effort started to increase because of the need for fish to feed the South African troops in this area during World War I. More important for the commercialization of the fisheries, was an increased demand for fish in the growing urban markets in Blantyre and Zomba as well as among workers on the tea plantations in the Shire Highlands in southern Malawi. In order to exploit these emerging markets, in the late 1920s and 1930s expatriate entrepreneurs from Europe and Asia began to invest – first in fish trade and later in fisheries – and this resulted in a substantial growth in

---

[4] Given the geographical proximity between the two water bodies, it is impossible to treat developments in Lake Malombe in isolation from what has happened on the South East arm of Lake Malawi.

fishing effort in the South East arm of Lake Malawi. With capital generated in various businesses, Indian and Greek traders invested in lorries and transport, and began to purchase fish from African fishermen. In addition to supplying the markets in southern Malawi some of the fish was also exported to Southern Rhodesia (Chirwa, 1996). After the 1930s, these entrepreneurs started to invest in their own fishery with large motorized vessels in the South East arm They also introduced new fishing gear like nylon nets, beach seines and trawls, which to some extent were to be taken up by the local population (Hara and Jul-Larsen, 2003). A growing number of African fishermen thus made a living by fishing for the growing urban markets, in addition to a group of expatriates with far more efficient gear, who also participated. Unlike the situation in Lake Kariba, the two fisheries on the South East arm competed for the same resources. The expatriate fishermen did not operate in Lake Malombe[5], but the African fishermen made sure that the new gear introduced in Lake Malawi was also introduced in Lake Malombe. On both lakes Africans and expatriates participated in fish trade although it was completely dominated by the latter group.

Many conflicts arose out of this situation, and the African leadership at the local level began to argue in favour of the right of African fishermen to own the natural resources. The British colonial authorities were also concerned about the ecological effects of the modern gear of the expatriates, and a number of restrictions were put on the expatriates while African fishermen were largely left to fish without too many restrictions. In the 1950s, a new group of African owners emerged. Instead of the traditional canoes, the new owners started to invest in planked vessels with engines and seines targeting the Chambo (*Oreochromis* spp.). They employed crews to fish for them and often operated both in Lakes Malombe and Lake Malawi. On the South East arm in particular, the number of small African entrepreneurs increased during the 1960s and the 1970s, but already in 1962, 30 of these African boat owners were registered to fish in Lake Malombe. The number of owners increased steadily throughout the 1960s and 1970s, and was over 200 by the early 1980s (Figure 3.5). On Lake Malawi this process was paralleled by a decline in the industrial fleet operated by expatriates. Of the several industrial units operating in the 1940s, only three remained in 1955. In the 1990s, only one company was operating in the South East arm.

Where did the capital that facilitated the investments among the African fishermen come from, and who were the investors? Though a limited number of new owners were initially supported with government loans, most of the capital that was pooled into the fisheries on both lakes came from work performed in other economic sectors (like shops or transport business) and particularly from remittances sent by Malawi migrant labourers in South Africa and Southern Rhodesia. Most of the new owners were migrants who returned after many years of working in the mines. This is how the Fisheries Officer in the area characterized the new owners in a Government hearing in 1956: "Since 1950 there have been a few Africans who have attempted to set up fisheries on a real business basis. (...) not one is originally a fisherman himself. They are all African businessmen (...) and have spent most of their lives in South Africa or Southern Rhodesia and have come back to set up business"[6]. However, even if many of the returned labour migrants invested in fisheries for the first time, a closer look at their often varied

---

[5] Between 1917 and the late 1930s, waters from Lake Malawi did not flow into the Shire river. As a consequence, Lake Malombe was small and varied considerably according to annual rainfall in its immediate catchment area (McCracken, 1987 and Mandala, 1990).
[6] A.D. Sanson's statement in: "Record of the meeting of the Commission of inquiry into the fishing industry held at the court house, Fort Johnston, 8th and 9th June 1956." (MNA/COM/9/3/1).

backgrounds, reveals that many of them actually had a background in fisheries and long experience from the fisheries of Lake Malawi and Lake Malombe (Hara and Jul-Larsen, 2003). Because of the influx of external capital, the number of fishermen at Lake Malombe remained fairly stable throughout the 1970s and 1980s, despite a noticeable change towards more effective fishing gear (see Figure 3.5). This shift towards more expensive fishing gear meant that the entry costs of the new fishery increased, and this implied that it became increasingly difficult for fishermen without capital to remain in the fishery. For these people, the only solution was to find employment among the crew of the owners. The increased input of capital and technology gave jobs to a rising number of workers, and to fish traders, who could participate in a marketing system where entry costs remained low: the investors in the fisheries seldom extended their activities to the marketing chain and they have never had any particular control over this activity.

In this situation, the number of owners stabilized in the 1980s, but their investments in boats and seines increased. By 1975, the catch rates of Chambo in Lake Malombe had been severely reduced. The boat owners managed to remain in the fishery by shifting their focus towards other species: the small-sized Haplochromine cichlids. This necessitated a change in fishing technology, first towards new beach seines (Kambuzi seines) in the 1970s, and later in the 1980s towards open water seines (Nkacha seines) (Figure 3.5).

During the 1990s, fishing effort in Lake Malombe went into a phase of recession. One of the explanations is certainly the collapse of the Chambo in the late 1980s and the reduced economic possibilities that this collapse entailed. Another important limiting factor for the maintenance of a capital-intensive fishery in Lake Malombe, was that labour migration to Zimbabwe and South Africa gradually came to an end when the mining sector in these countries entered into crises in the 1980s and the main source of capital for investment in Lake Malombe fisheries dried up. In combination with generally hard times in most other sectors, especially in the 1990s when the Malawian economy was liberalized and restructured, other sources of capital were also meagre.

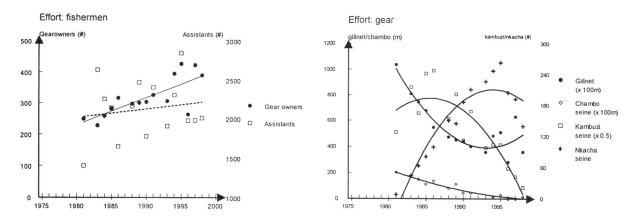

Figure 3.5 *Effort development: number of fishermen and gear in Lake Malombe.*
Source: Zwieten and Njaya, 2003.

3.3 Trends in effort dynamics: Population-driven rather than investment-driven growth

The above descriptions of effort development in three lakes serve to illustrate the considerable variability in the growth of fishing effort that exists in the SADC area. This variability can be observed both geographically and historically. With regards to geographical variations, one lake can experience strong growth in fishing effort while others do not experience much growth at all, such as on the Zambian side of Lake Kariba where most of the long-term growth is linked to the introduction of the Kapenta fishery. Effort development on the Zimbabwean side of Lake Kariba shows a very similar pattern to the one just shown for the Zambian part (Kolding *et al.*, 2003a). Differences in growth trends between lakes often seem to emerge despite clear similarities in the ecological and structural factors conditioning the fisheries. With regards to variability over time, one can observe that whereas one period may be characterized by strong acceleration in the growth of effort, other periods are calmer and may even experience reductions. We saw this in the Zambian part of Lake Kariba as well as in Lake Malombe, but the same is also to some extent the case in Lake Chilwa (see Chapter 5). Finally, the variability may be expressed in a combination of the two types of differences: increased growth in one water body seldom seems to take place during the same period of time as in another water body.

The cases thus serve to illustrate how problematic it is to consider growth in fishing effort as empirically inevitable. They remind us of the existence of multitudes of natural and social factors that in various ways and to various degrees influence the millions and millions of individual micro-decisions taken daily by fishermen about if, how, when and where to fish, and which on the aggregate level constitute the changes in fishing effort. Sometimes, these factors may affect the fishing effort in radically different ways, even when demographic and economic conditions seem to remain fairly stable, as is the case in the shift from high growth in the 1980s to stabilization and regression in the 1990s in the Zambian inshore fisheries of Lake Kariba. Assumptions about effort development based on common property theory seem therefore to be of limited help in interpreting fisheries development in the SADC freshwater bodies.

However, despite the considerable empirical variations, certain aggregated trends in the effort development can be seen when we introduce the distinction between population- and investment-driven growth in fishing effort. This enhances our ability to discover under which circumstances the theory is of limited empirical interest and it also helps us to formulate the right questions about why the assumptions do not apply (Brox, 1990). Although most of the fisheries in question have experienced elements of both types of growth, the population-driven growth has been and still remains by far the dominant feature. In Lakes Chilwa, Chiuta and Mweru Wa Ntipa, as well as in the main rivers and floodplains like Bangweulu and Okavango, fluctuations in fishing effort over the last 50 years have been almost exclusively related to the number of harvesters, whereas very little has been caused by more efficient technology or fishermen's accumulation of gear. Fishing methods, species targeted, volume of gear per unit and the organization of production have been relatively stable over time, even if they may have fluctuated seasonally or periodically according to changes in the natural environment. Both Figures 3.2 and 3.3 illustrate this and demonstrate high stability in types of gear over time and a strong correlation between numbers of gear and numbers of fishermen.

By contrast, investment-driven growth does not seem to have played an equally important role in SADC freshwater fisheries development. In principle, we may distinguish between two types of investment-driven changes: either investment derives from new groups of people from outside with more efficient types of gear who manage to establish themselves as harvesters in an area, or the local fishers themselves manage to accumulate and develop their existing gear. In the SADC area, there has been a general interest among certain groups of foreigners in investing in many freshwater fisheries. Especially in the big lakes, but also in many of the intermediate and smaller ones, we have seen that expatriate entrepreneurs at different points in time have been ready to invest in various types of mechanized fisheries. Before the 1960s, investors were often of southern European or Asian origin, but in later years there have also been many South Africans, as well as some initiation of investments by African entrepreneurs. In general however, most of these investment-driven changes have not been very successful. Even if some of the attempts have lasted for many decades, the great majority of them have failed. In Lake Mweru for example, the mechanized fishery of the foreigners that emerged in the 1940s and which eradicated the Mpumbu has been non-existent for many decades. In Lake Malawi only one mechanized fishing company of foreign origin has survived and in Lake Tanganyika it is only along the very short Zambian shoreline at the southern end that the industrial fishery which used to operate all around the lake, has survived (Magnet *et al.*, 2000). In Zambia the mechanized Kapenta fishery on Lake Kariba survives, although it is not expanding at present and in recent years it has even shown signs of decline (Overå, 2003).

Investment-driven growth initiated from inside the fishing communities is even more difficult to observe. From a historical perspective, the most important examples may be the switch from the use of locally produced fishing equipment to the utilization of manufactured fishing gear. According to studies undertaken in other African freshwater fisheries, this change led to a considerable increase in fishing effort because more time (previously used for production of gear) could be allocated to production purposes (Quensiere, 1994). This change, which in most cases took place in the 1940s and 1950s, was general and has affected virtually all fishermen. However, it does not mean that locally produced gear has disappeared: a range of various home-made gear is still effectively utilized. These gears may be used in combination with gillnets or seines, which have become common means of production among the great majority of fishermen. However, the shift towards manufactured gear has not resulted in a continued process towards appropriation by particular individuals or groups of increasingly effective catch equipment. Only in the big lakes do we observe the slight emergence of groups of fishers where commercial production has visibly started to influence social organization and behaviour in some of the same ways as the West African coastal canoe fishery has become so famous for (Chauveau and Jul-Larsen, 2000). Only in Lake Malombe, which can be considered an extension of Lake Malawi and where fishermen often fish in both lakes, have we seen investment-driven changes in fishing effort where the investments have derived from the local population. But, as Fisheries Officer A.D. Sanson observed in 1956 (see previous section), the capital had not been generated locally but was accumulated while abroad on work migrations.

Differentiating between population- and investment-driven changes in effort proves to be more complicated than it first appears. Technological change towards more effective gear is defined by Brox as an expression of 'vertical' (or investment-driven) growth of effort (1990: 233). In the case of Lake Mweru we observed different processes of technological change, but to classify these changes as investment-driven may be too simplistic. Gordon (2002) argues that the technological changes observed in the 1950s, when foreign actors invested heavily in fishing technology as a response to the emergence of new markets for fish in the Copperbelt, undoubtedly must be considered as investment-driven changes. But the new technologies due to the introduction of Chisense fishing in the late 1970s and in the 1980s were a different kind of process. According to Gordon, these technological changes must be mainly considered as responses to diminishing catches of certain species, which in turn were caused by an increase in the number of fishers. Even if the new technologies to some extent implied a minor increase in financial investments, they were not introduced as a result of changes in the investment patterns among the fishermen. The changes were, according to Gordon, necessary simply in order to be able to remain a fisherman, and did not result in an increased efficiency of fish production per se. He therefore concludes that even if this type of growth in effort often results in technological changes, the changes must mainly be classified as "horizontal" or population-driven changes.

Gordon's interpretation also has implications for the interpretation of the technological development in Lake Malombe. There, the introduction of Chambo seines in the 1950s and 1960s was clearly driven by changes in investment patterns as a result of the return of the international work migrants. This was also the case with individual accumulation of various types of fishing gear and other technological changes (e.g. introduction of planked boats and outboard engines) up till the beginning of the 1990s. There have, however, been other technological changes in the Lake Malombe fisheries that could be considered responses to a reduced availability of certain species (in this case Kambuzi). In that respect there is little difference between the replacement of gear targeting Chambo with gear targeting Kambuzi in Lake Malombe (Kambuzi seines and Nkacha nets). These examples of technological changes would thus, by Gordon's interpretation, be regarded as population-driven changes of effort, rather than investment-driven technological expansion.

With regard to Gordon's argument, some empirical questions may be raised: firstly the fact that the Chisense fishery has never substituted, but rather has supplemented the gillnet fishery (Figure 3.3). Furthermore, the figure also shows how the Chisense seines have never become as numerically important as gillnets. The empirical side of Gordon's argument is therefore not without problems, but in general terms we think it remains valid: technological change is not necessarily an expression of investment-driven changes in effort. Fisheries development in the area of this study is full of examples of local technological improvements that cannot be defined as investment-driven.

Nevertheless, we will stick to Brox's definition and maintain that most technological changes that imply significant changes in selectivity and effectiveness (see Chapter 5) – which until now have been relatively rare in SADC freshwater bodies – must be defined investment-driven. Except for some cases of foreign investment (which have often failed) and the introduction of manufactured gear, there are hardly any examples in our data where fishing effort has developed due to investment-driven growth. The obvious exception is Lake Malombe where much of the technological changes and the accumulation of gear among African fishermen after 1950 can only be understood as investment-driven changes. Lake Malombe is thus a case where the investment-driven changes emerge as a combination of local and foreign inputs in the sense that many of the investors were "insiders" from the Malombe area, but they derived their means of investment from other economic sectors and geographical areas. Similar processes of investment-driven growth are also observed in the other great Eastern African lakes. The rapid development of an export-oriented Nile Perch business in Lake Victoria in the 1990s happened as a result of big investments in processing technology, mainly by wealthy segments in the bordering states (Jansen, 1997; Abila and Jansen, 1997). Although most of the immediate investments were channelled into processing, they also entailed investment-driven growth of fishing effort (Mbuga *et al.*, 1998). However, as the evidence from our case studies of investment-driven growth has illustrated, it is still too early to know to what extent the changes in Lake Victoria will be long-lasting. On Lake Tanganyika we have seen a less dramatic type of investment-driven growth. Not unlike what has happened in the West African marine fisheries (Chaboud and Charles-Dominique, 1991), a part of the small-scale fishery on Lake Tanganyika changed into a more dynamic and more capital-intensive type of production based on lift nets in the 1970s (Coenen, 1994; Paffen and Lyimo, 1996). As in the case of West Africa (Haakonsen, 1992) it seems as if this more capital-intensive artisanal fishery has been able to a large extent to out-compete the industrial fisheries that have been established there since in the 1940s.

To summarize this discussion, one may say that at an overall level population-driven growth in effort has been far more predominant in SADC freshwaters over the last 50 years than investment-driven growth. Among the water bodies included in this study, Lake Malombe is the only case where investment-driven changes in fishing effort seem to have been the dominating trend; in the others it is the population-driven changes that constitute the most important element in effort dynamics. The fisheries in this region thus remain technologically simple and accumulation of gear within the units is limited. Interestingly, this situation is very similar to what Brox argues to have been the case in the northern Norwegian fisheries until the 1930s.

3.4 Mobility among fishermen prevents fishing from becoming a 'last resort'

Literature on fisheries management rarely distinguishes between population- and investment-driven growth in fishing effort. One reason for this omission may be that in principle, one has tended to consider the biological and economic effects to be the same (for the fish stocks), independent of whether the growth is caused by population increase or investments. Nevertheless, there are some scholars who have addressed this issue. During the last decade, Daniel Pauly is one of the most influential contributors who indirectly distinguish between the causes for change in effort in his analysis. Through concepts like "Malthusian overfishing" and fisheries as a 'last resort' Pauly (1994; 1997) argues that even if the production methods in many small-scale fisheries hardly change (i.e. are only modestly subject to investment-driven changes), an uncontrolled entry by poor people who have been marginalized from other economic sectors – in particular from agriculture – will lead to human overpopulation and to overexploitation of the biological resources. This 'last resort' situation resulting in a heavy influx and concentration of people from other occupations in the fisheries is, according to Pauly, caused by people's loss of access to other vital resources (such as land or animals). They simply have no alternative means of livelihoods: "... the most worrisome development within the small-scale fisheries of tropical developing countries in Asia, Africa and Latin America is the entry of non-traditional fishers into the these fisheries .... In all cases, these people enter fisheries because they have been forced out of their traditional occupations, because there is excessive pressure for land, or because lack of access to grazing range has marginalised livestock production in inland areas" (Pauly 1997:42). Pauly also shows that the distinction between population-driven and investment-driven growth in effort becomes important when we want to deal with remedies and solutions to the problems: "... the concept [Malthusian overfishing] was developed, and the term coined, to stress that coastal systems cannot continue to serve as a convenient dump for excess labour and produce ever increasing or even sustained amount of goods and services" (1994: 117). According to this view, poor people's access to the resources must be controlled in order to prevent them from causing an ecological and economic 'tragedy'.

We have already shown that the development in the SADC freshwater bodies supports Pauly's emphasis on the importance of population-driven growth of fishing effort. However, it becomes more complicated when we investigate the assumption of fisheries as a 'last resort'. Pauly insists on a high demographic mobility in fisheries, but in his analysis this mobility mainly goes in one direction: into the fisheries. He assumes (at least implicitly) that because of the loss of access to vital resources in other sectors, a fisher tends to remain a fisher once he has become one. For Pauly, small-scale fishing in the developing countries thus represents an occupation of 'no return'. This interpretation is difficult to substantiate in the case of SADC freshwater bodies. In our studies, we have observed very few cases of populations that may be called specialized fishers or fishing communities. Rather, the producers combine fishing with other occupations – in most cases with agriculture, but also with other activities considered as interesting income-generating opportunities. In this regard, fishermen in the SADC region are not different from African small-scale fishermen and farmers in general (Chauveau and Jul-Larsen, 2000): they try, as much as they can, to diversify opportunities and risks between many economic activities. Indeed, the majority of the fishers in this study combine fishing in parallel with other economic activities – simultaneously, seasonally, or sequentially.

Our studies emphasize the importance of access to land, even for the fishers. When the non-Tonga immigrant fishers on the Zambian side of Lake Kariba experience difficulties in combining fishing with agriculture and/or animal husbandry due to problems of access to land (see Box 4.1), this causes severe problems. Our surveys show that many of them are able to practise other occupations (like petty trade and various artisanal activities) in periods of reduced productivity in fisheries (Jul-Larsen, 2002), but land always remains the most important asset. Our studies also expose the way in which immigrants in the long run are slowly able to overcome the problems of access to land. Through marriage relations and other types of alliances (e.g. foster children), by pushing for burial grounds for their dead, and through the management of witchcraft allegations, and so on, immigrant fishers acquire access to land and other resources that enable them to diversify their sources of income beyond the fisheries. In the case of Lake Kariba, as in the SADC region and in the rural sub-Saharan Africa in general, it is therefore still very uncommon to observe fishermen being completely deprived of land – unlike what may be the case in other parts of the world[7].

The mobility between livelihood sectors is not only seen in the extent to which people combine and pursue parallel occupations. It is also expressed in people's shifting in their main economic activities over time. For example, our Lake Kariba survey shows that more than 70% (out of 426 fishermen) explain that they had other occupations before they started fishing on Lake Kariba (Jul-Larsen, 2002). Frame surveys in Lake Mweru show similar figures (Zwieten et al., 2003b). As these fisheries are dominated by simple and capital-extensive technologies, entry fees are low, and this facilitates the mobility of people into the fisheries. Also in terms of knowledge and organization, the simple level of the technologies favours easy entry for those who wish. This means that in the case of Lake Kariba, practically anyone can become an economically independent inshore fisher within a few years. From the 'last resort' perspective, the danger of overpopulation and overexploitation is inevitable.

However, according to our observations, people are just as likely to leave the fisheries as they are to enter it. In the case of Lake Mweru we saw (see sub-section with same name, earlier in this chapter) how about 3 000 fishermen left the fisheries between 1992 and 1997, which at the same time was a period characterized by significant net growth in the number of fishermen. In Lake Kariba, there have been two periods (since the creation of the lake) of substantial reduction in fishing effort; the years immediately after 1963 and in the 1990s. Our surveys also show that, during the 1990s, more people left the fisheries compared to the net reduction in fishing effort.

---

[7] Note the fact that Pauly has had most of his small-scale fisheries experience in Asia and that this may explain his viewpoints.

Our quantitative data only indicate that people have left a particular fishery and not that they have left fishery as an occupation. This is due to the general difficulty in providing exact data on people's departures from the sector, since this would require extensive collection of data in places outside the typical fisheries setting (the boat, the landing site, the markets or the local village), and it is often difficult to locate where to look for such data. However, a range of life stories and interviews with family members still remaining in the fisheries in all of our study areas, indicate that people leave the fisheries as easily as they join them and that opportunities are thought of and actively sought outside as much as inside the sector. The causes of and the mechanisms behind this mobility are some of the issues we will look at in Chapter 4. However, before ending this chapter we will briefly consider to what extent central fisheries management practices in the region have influenced the trends in fishing effort just described.

3.5 Effort development in the context of the fisheries management history

To what extent has the effort development outlined in this chapter been influenced by central management practices? Conservation of natural resources already has a long history in Southern Africa. Explicit natural resources management emerged in the first part of the nineteenth century as a consequence of general colonial policies, and an extensive literature on the experiences of this began to appear in the late 1980s (Anderson and Grove, 1987; Grove, 1988 and Beinart, 1989). However, it is fairly recent that fisheries research has started to include the management history in its analysis (see e.g. Chirwa, 1996). Two recent doctoral theses analyse the role of fisheries management in Malawi (Hara, 2000), Zambia and Zimbabwe (Malasha, forthcoming). A specific case study, based upon Malasha's thesis has been prepared for this study (Malasha, 2003). These works tend to reinforce many of the results in the research from other productive sectors, which emphasize two aspects that are particularly important for an understanding of what the legislation and the practical regulations came to look like – one ideological and one political.

The ideological aspect is related to how regulation measures came to reflect the prevailing representations and images among the colonial administrators of the relationship between man and nature in general and the relationship between Europeans/Europe and Africans/Africa in particular. As emphasized by Anderson and Grove, the image of "... Africa [as opposed to Europe] as a wilderness in which European man sought to rediscover a lost harmony with nature and natural environment" (1987:4) has, since the nineteenth century, been a strong underlying premise for the formulation of conservation policies as well as the creation of practical regulations. According to the same authors, the natural environment has been considered as "... a special kind of 'Eden', created for the purpose of the European psyche, rather than as a complex and changing environment in which people have actually had to live. The desire to maintain and preserve 'Eden' has been particularly pronounced in eastern and southern Africa ..." (Ibid.). No wonder, then, that the desire to conserve nature sometimes became more important than the wish to investigate the complex and changing environment.

Malasha (forthcoming and 2002) shows how Dr. C. F. Hickling, a fisheries biologist in the Colonial Office in London, already in 1952 emphasized the problems of utilizing conventional game conservation models in the management of tropical fisheries. In a five-page 'Memorandum on Fisheries Legislation' he concludes that most of the restriction and prohibitions used to regulate the fisheries in the British colonies and overseas territories since the early twentieth century have probably had very limited effects: "... fisheries legislation, in so far as it aims at the conservation of fish stocks, is not a simple matter, but a highly complex one, in which the results of regulations may not only differ from those intended but may even defeat them"[8]. In Hickling's view, the licensing of gear or nets that requires large and expensive enforcement staff, and other measures such as closed seasons, mesh-size regulations, and fish-size regulations are not very useful either. Nevertheless, and despite Hickling's strong reservations about most of the applied management measures, they have continued to form the basis of freshwater fisheries management in most countries in Southern Africa until today.

The second aspect that is highlighted in the literature is of a more direct political order and concerns the distribution of power and the pursuit of interests among the hegemonic groups. In order to suit and to serve the overall purposes of the colonial powers, the exploitation of fish, just like any other resource, had to be defined according to their interests, and fishers had to be controlled for the same purpose. But the colonial policies could vary considerably and the management regulations would do the same. As shown in Malasha's case studies (2002), the differences in fisheries regulations on Lake Kariba between what is now Zambia and Zimbabwe can to a large extent be seen as a reflection of the different political interests between Britain's colonial administration that dominated in Zambia (Northern Rhodesia) and the settlers, who had a strong influence on how laws and regulations were formulated and practised in Zimbabwe (Southern Rhodesia). In the case of Zambia, the development of a sound mining industry in the Copperbelt and the securing of cheap food for its labour force were very important for the British colonial authorities, and these interests influenced the guidelines that they developed for the regulation of fishing in that country. Similar concerns were to some extent also the case in Malawi (Hara, 2000). In Zimbabwe, on the other hand, fisheries regulations rather reflect the concern for the rights and privileges of the white settlers, who considered freshwater fishing as a leisure activity. The angling interests thus became paramount in the formulation of fisheries and stocking policies. One can still see how the only function of certain mesh size and locality regulations in the Zimbabwean part of Lake Kariba is to conserve the large specimens of the Tigerfish (*Hydrocynus vittatus*), which were the preferred target for anglers (i.e. settlers) on the lake. In later years the angling interests have been taken over by the tourist industry which, through its economic importance and political influence, has been able to uphold many of the fishing regulations.

The legacy of the colonial conservation practices is probably much stronger than is usually imagined. Besides, the present central management regimes also reflect new ideological images and the interest of new groups with influence in the independent states in question. Despite the fact that policies often emphasize the need to limit effort, it is interesting to note that only Zimbabwe has licence regulations in place which can effectively limit population-driven growth in effort (see Table 3.3). All the other measures regulate effort rather than limit it.

Despite the lack of surveillance the central management regulations have certainly affected the life of fishers and production at the local level. The lack of control does not mean that some groups may use the regulations against other groups. However, given that central regulations generally do not include limitations on the number of producers, we may conclude that the central management regimes probably have limited effects on the growth in fishing effort. In Chapter 4 we shall seek for the causes elsewhere.

---

[8] Ref. Footnote 1, Chapter one.

Table 3.3 *Different management measures applied in five fishing areas.*

Management measures	Mweru (Zambia)	Kariba (Zambia)	Kariba (Zimbabwe)	Malombe (Malawi)	Chilwa (Malawi)
Licences					
- number of fishers			x		
- number of gear			x		
Restriction on fishing methods					
- type of gear	x	x	x	x	x
- mesh size	x	x	x	x	x
Time and spatial regulations					
- closed season	x			x	
- closed or protected area	x	x	x	x	

4. FACTORS BEHIND CHANGES IN FISHING EFFORT

4.1 Introduction

Although common property theory (CPT) does not directly include assumptions about demographic growth, there is a very strong tradition in the management literature to do so. Pauly's attempt (1994) to restore some of the main arguments developed by Thomas Malthus 200 years ago is certainly not the only example. Garret Hardin was also strongly influenced by Malthusian ideas in his famous article from 1968 where he "clearly connects the tragedy of the commons to over-population", as Brox (1990:233) puts it. Our own findings on how fishing effort has developed in SADC freshwaters actually support the assumption that demographic growth also entails increased fishing effort.

Table 4.1 *Demographic growth in the economically active population of 12 SADC countries (in thousands).*

	1970	1980	1990	1970-90 (%)
Angola	2 895	3 472	4 327	49.5
Botswana	286	395	548	91.6
Dem. Congo	9 512	12 005	15 880	66.9
Malawi	2 360	3 112	4 578	94.0
Mozambique	5 258	6 686	7 520	43.0
Namibia	360	446	561	55.8
South Africa	8 249	10 568	13 526	64.0
Swaziland	158	200	254	60.8
Zambia	1 898	2 399	2 953	55.6
Zimbabwe	2 386	3 201	4 574	91.7
Total	41 152	52 557	68 534	66.5

Source: FAO, 2000b

However, as shown in the previous chapter, population-driven changes in effort reflect far more complex social processes than simple birth and death rates. If one takes a look at the development in the economically active population between 1970 and 1990 in the 12 SADC countries presented in Table 4.1, one realizes that demographic growth[9] only provides a partial explanation. While the economically active population increased by 67 percent between 1970 and 1990, according to Table 3.2 the number of fishermen increased by about 160 percent during approximately the same period. Furthermore, demographic growth cannot explain the dramatic shifts in effort that have been observed in the fisheries of Lake Kariba, where the number of fishers decreased by 75% in less than four years after 1963, and then increased by 150% in seven years during the 1980s. Lake Chilwa is another example where demography fails to explain growth in the number of fishers. Fishing stopped completely when Lake Chilwa dried up in 1994, but during the four consecutive years the number of fishers nevertheless grew by 55% compared to what it had been just before the drought. Neither can demography explain the exceptional stability in the number of gear owners and assistants after 1980 in Lake Malombe; situated as it is in an area where the demographic growth is one of the highest in Africa.

The first question that this chapter will address is therefore: which factors other than demographic growth are important for the understanding of population-driven growth in effort? The second question concerns factors that stimulate – or hinder – investment-driven growth in effort. Proponents of CPT assume that fishing effort grows as a result of increased investments. Based on the analysis of a prisoner's dilemma situation, free access to the resource and the rent it represents, are considered to be the driving forces behind continually increasing investments in production material until the rent has been completely deleted. This becomes a vicious circle between investments, fishing effort and catches: increased investments lead to increased effort, which results in decreasing catch rates, which in turn leads to further increase in effort, and so on. But in Chapter 3, one of our main findings was that investment-driven growth has been very modest in the SADC freshwater fisheries over the last 50 years, irrespective of whether the investors have represented local fishermen or have been expatriate entrepreneurs. In the case of the former, investments have never lasted long enough to result in lasting increases in the investment patterns. Local fishers do not tend to accumulate fishing gear and other means of production; the work organization remains simple with few people involved; and production units tend to remain based on relations of family and kinship. Although fishing technologies have often changed, such changes are frequently necessitated by changes in the environment or by increases in the number of fishermen, rather than being an expression of the producers' search for a greater share of the resource rent.

Of the four lakes directly included in this study, only Lake Malombe shows major investment-driven changes in effort initiated by the local population. With regards to foreign investors, we have seen many attempts to introduce more capital-intensive production systems. In fact, many Europeans (mainly from the Mediterranean countries) and Asians have tried to make a living in African freshwater fisheries throughout most of the twentieth century. In the last 20 or 30 years, an increasing number of African entrepreneurs have also tried. But these attempts have tended to fail, and in most lakes in the SADC region one may find idle remnants of catch and processing technology brought in by foreign entrepreneurs. In some cases, as in the Kapenta industry in Lake Kariba, the producers are still active even if the levels of investments – particularly on the Zambian side – seem to have fallen during recent years.

Hence, the extent to which we should continue to regard CPT as a useful analytical tool, depends on the answer to the third question that this chapter deals with: why are the CPT assumptions with regards to effort development not valid in SADC freshwaters – why do fishers in these

---

[9] For a definition of demographic growth, see Chapter 2.

areas not increase their fishing effort by increasing their investments? Let us first take a closer look at the causes behind the population-driven changes in effort.

4.2 Factors influencing population-driven growth in effort

Since fishing effort basically is an aggregate result of an infinite number of micro-decisions taken on a daily basis by the individual fishers or "fishers-to-be", one may say that there are few aspects in peoples' daily lives that do *not* influence their decisions on whether and how they go fishing. However, what we wish to highlight here are factors that can be considered to be of considerable importance for effort development in SADC freshwaters, and that national authorities therefore should take into account in the formulation of management policies. In broad terms, the data of our project indicate that in addition to demographic growth, at least four factors have considerable influence upon population-driven changes in the freshwater fisheries: the natural productivity in each lake; alternative livelihood opportunities; local access-regulating mechanisms; and increased capital requirements in the fishery.

4.2.1 Natural productivity and macroeconomic changes

One of the most significant characteristics of SADC freshwater ecosystems is the large environmental variation that is manifested in fluctuating water levels (see Chapter 5). Research findings from African production systems based on natural resources other than fish show similar fluctuations in the productivity of the ecosystems due to considerable climatic variation (Scoones, 1995). With regards to freshwater fisheries, the most obvious examples are shallow (but potentially very productive) lakes, such as Chilwa, Chiuta, Malombe and Mweru Wa Ntipa, where the water disappears from time to time. Lake Chilwa dried up completely both in 1969 and 1994. And for 20 years (1917-1937) the water connection between Lake Malawi and Lake Malombe was cut off, and Lake Malombe dried up on several occasions during this period (McCracken, 1987). In the first half of the 1990s, water levels in the Shire River reduced considerably again. Also in more permanent lakes such as Lake Mweru and Lake Kariba, there is considerable variation in water levels. Increased water levels generally entail improved inflows of nutrients, which normally bring an increase in the ecological productivity of the system. Such increases in productivity can be considerable and lead to booms in the fisheries of certain species.[10]

Changes in the natural productivity of the lakes seem to influence the population-driven changes in fishing effort. Increased productivity tends to attract newcomers while reduced productivity clearly has the opposite effect. It is not difficult to imagine how the drying up of shallow lakes also leads to reductions in the number of producers. The opposite happens in periods when inflows (and water nutrients) are high, which leads to an increase in the recruitment of fishermen. This was the case during the filling up of Lake Chilwa after 1994 and in Lake Mweru Wa Ntipa in the 1980s. In Zambia, where access to the new fishery in Lake Kariba was relatively open for people from the Zambezi valley, recruitment rates were very high during the years of inundation. Immediately after the lake had filled up (in 1962-1963) there was a sudden reduction in productivity, which was followed by a dramatic reduction in the number of fishermen.

Fluctuations in the number of producers according to changes in ecological productivity are also reinforced by their high geographical mobility. For example, it is reported that some of the

---

[10] The enormous increase of *Tilapia* spp. in lake Turkana in 1976 is an example of such a boom (Kolding, 1989).

fishermen on Lake Chilwa go to Lake Malombe and Lake Malawi to work during periods when water levels fall. Also, the increase in effort in Mweru Wa Ntipa in the 1980s was a result of heavy immigration to the area (Skjønsberg, 1992). A similar kind of mobility appears in our surveys from Lake Kariba: many of the persons interviewed report that they had been fishing in other water bodies such as in the Kafue plains before they arrived at Lake Kariba.

Fluctuating productivity in the different water bodies affects peoples' opportunities and thereby their decisions with regards to fish production. These decisions can therefore not be considered in isolation, but must be seen in the context of factors that change their opportunities in other livelihood sectors. In Chapter 3 it was emphasized that occupational mobility among SADC freshwater fishermen is considerable, and that this is mainly expressed through different combinations of fisheries and agricultural activities. Fishing is also combined with other occupations; in particular, wealthier fishermen tend to combine many sources of income. As has been shown for Lake Malombe and the South East arm of Lake Malawi (Hara and Jul-Larsen, 2003), most of the gear owners combine their activities in the fisheries with also being involved as shopkeepers, transporters, bar owners, flour mill managers, carpenters, charcoal producers, etc. Changes in the opportunities in parallel sectors are therefore as important for the development of fishing effort as changes in the natural productivity of the water bodies. Unlike what has been observed in other parts of Africa (e.g. Chauveau and Jul-Larsen, 2000), professional specialization in small-scale fisheries in the SADC region is not very common, and there are few mechanisms whereby professional rights or identities are ascribed to particular groups of people (as for instance in the form of a caste system which one finds among fishermen in Mali, ref. Quensière, 1994). Whereas professional stability in many parts of the world is reproduced through the control of knowledge, the great differences in terms of knowledge and experience between various groups of fishermen in SADC, and the relatively simple levels of techniques and technologies, tends to facilitate professional mobility instead.

In addition to many combinations of occupations, there are also many ways to organize them. The same person may undertake fishing and other activities simultaneously, or sequentially. Furthermore, the combination may be based on some sort of specialization between individuals within the same production unit, or it may be a mixture of different modes of organization. It therefore becomes difficult to identify aggregated patterns and to generalize about the extent to which developments in parallel sectors influence effort development in fisheries. One important exception seems to be the role of what is often referred to as the modern or formal sector; i.e. salary employment in major private or state-owned enterprises. Contrary to what is often believed, there are close connections between formal employment opportunities and informal income-generation opportunities, like fishing. Fluctuations in the macro-economic conditions therefore directly affect fishing effort to the extent that they affect job opportunities.

Gordon's case study from Lake Mweru (Gordon, 2003) shows that it was mainly people who had lost their jobs in the mines of the Zambian and Zairian Copperbelt after the economic crisis that started around 1973-74, who came to introduce the new Chisense fishery. Similarly, the considerable growth in number of fishers on the Zambian side of Lake Kariba in the early 1980s is explained by the same phenomenon. People from all over the country (but particularly from the north) who had lost their jobs – first in the Copperbelt and later in the state sector in urban areas throughout the 1970s – immediately came to join the Lake Kariba fisheries when the Zimbabwean War of Independence ended in 1980. In a survey that included 250 "foreigners" who arrived after 1980, 83 percent reported that they came from jobs in the Copperbelt or in Lusaka (Jul-Larsen, 2002). Not all of them had jobs in the formal sector; many also worked in more informal enterprises which depended on the formal mining economy. Another example of the influence of macro-economic conditions on fishing effort is Lake Chilwa in Malawi, where the extraordinary increase in number of fishermen after the refilling of the lake in 1995 was partly explained by the return of labour migrants, who had been hit by the loss of opportunities in the South African mining economy.

There is no doubt that the overall economic crisis which has troubled most of the SADC countries for the last 20 to 30 years is one of the main reasons behind the very high increase in numbers of fishermen. However, since the crisis has lasted for many decades and has hit all the countries, it is difficult to assess the extent to which an economic recovery would reduce the number of producers. There are a few cases where employment opportunities in the formal sector have improved. They occurred a long time ago. One example concerns the dramatic reduction in numbers of fishers on the Zambian side of Lake Kariba in 1963. Although it must be considered a reflection of changes in the ecological productivity of the lake, as we described above, the reduction was probably strengthened by relatively good job opportunities in the formal sector during the years just after independence. The Gwembe Tonga already had good knowledge and experience from work migration to other parts of the country as well as to Zimbabwe, and thus easily gave up fishing when opportunities elsewhere improved.

Similarly, the lack of increase of fishermen in the Zimbabwean part of Lake Kariba as long as macro-economic conditions were good there, as described in Kolding's case study (Kolding et al. 2002a), may be another indication of the "preventive" effect of alternative job opportunities on population-driven growth.[11] Besides, interviews with hundreds of fishermen leave little uncertainty as to what fishers would choose if they could: jobs in the formal sector are far more attractive to them than their present situation as fishers.

4.2.2 Local access-regulating mechanisms

Neither the heavy increase in the number of producers, nor the high mobility between occupations in the fisheries and in other sectors, should be interpreted to indicate that people's access to fisheries resources is "free" or "open". However, in comparison with situations where mechanisms of ascribed rights, knowledge or other mechanisms directly limit people's movement between occupations and access to resources, people in SADC freshwater fisheries compete for such rights and access on a broader basis. In these societies, competition for access to fisheries is generally high, and there is always a range of social institutions that regulate and sometimes limit people's access to the resources. With regards to such institutions, it is important to reconsider the distinction between free access and common ownership that was established by the critique of classical CPT. The fact that people living around a lake have more or less equal rights to its waters does not mean that people from outside this local area can claim the same rights. In most villages or camps there are mechanisms that limit newcomers' possibilities of settling in the community – a precondition for the allocation of rights to natural resources.

In some cases, rights to the commons are directly regulated through the residential rights; if one is allowed to live somewhere, one is also allowed to exploit the shared resources. However,

---

[11] The lack of growth of numbers of fishermen in Zimbabwe must also be seen as a result of the land tenure system and government regulations along the Zimbabwean shores. This may explain why the number of fishermen does not seem to have increased during recent years of severe deterioration of the national economy.

given people's high geographical mobility, it is often problematic to define the physical and social "boundaries" of a community and people's "belonging" to this particular locality. Residential rights for outsiders or newcomers may also be associated with a particular activity (e.g. trade), which does not automatically give the individual shared rights to exploit resources. Furthermore, common rights based upon an ethnic or regional identity have tended to be in conflict with government policies – particularly after the countries achieved political independence. At the local level, one therefore often finds a range of unexpected mechanisms that aim to exclude residents from rights to common resources. It is for example puzzling that so few "foreigners" seem to have invested in the Lake Malombe fisheries. Despite their noticeable presence in the villages around the lake, we did not discover mechanisms that openly excluded them from participation. However, when we interviewed those few who had tried to invest in the fisheries, they told us how not only the other fishermen, but also the chiefs, would prevent the foreign boat-owner from dealing directly with his own crew whereby he was constrained in the management and decision-making of his unit. A foreigner is therefore forced to appoint a local representative to run the business on his behalf. Symbolically, this is expressed by a prohibition (a taboo) imposed by the local residents upon the foreigner against visiting the lakeshore when his boat returns from the lake. Not surprisingly, then, the few units belonging to foreigners were among those that experienced the biggest economic management problems. The deprivation of access of "foreigners" to resources is an important factor in the understanding of why the number of fishers has been relatively stable in Lake Malombe since 1980 (see Figure 3.5).

As mentioned above, the occupational mobility of fishers leads to a situation where regulations in the access to other natural resources, and to financial resources as well, may have a direct influence on effort development in fisheries. This was the case on the Zambian side of Lake Kariba in the 1990s. In order to illustrate how complex this process is and how many actors may be involved, we present this case in Box 4.1. It also explains why and how the number of fishermen, according to official frame survey data, declined by 50% during the 1990s (see Figure 3.4).

Box 4.1 *Access regulation and reduction of fishermen in the Zambian part of Lake Kariba*

Since the beginning of the 1990s, the Zambian Department of Fisheries (DoF) – together with their Zimbabwean counterparts – have run a fisheries management project on Lake Kariba. In 1994, the DoF presented what they called a co-management plan for the inshore fisheries. The key element of the plan was to relocate the "foreign fishers" (who had specialized in a very mobile adaptation by fishing from many different locations) into a few new permanent camps. Each camp was to be given exclusive and common rights to a defined zone of the shoreline. The definition of regulations in the zones was supposed to be "participatory" and involve both the fishers and the DoF, and the enforcement was to be left in the hands of the producers (Chipungu and Moinuddin, 1994). In addition to the inshore fishermen, a range of local actors such as traders, Kapenta operators, traditional chiefs and local government were invited to participate in the process. The co-management system was supposed to be funded through levies on traded fish.

The Kapenta operators were far too vulnerable to operate alone. They needed to ally themselves with other groups and it turned out that the traditional chiefs had equally strong (but different) interests in pushing the initiation of the management plan. The number of fishermen from outside the Zambezi valley had increased very strongly in the 1980s as a result of lost job opportunities in the formal sector. This caused a series of problems along the lakeshore, which in fact were more related to the access to land than to fishing grounds. The local population were willing to give the newcomers access to the fisheries. However, when the newcomers also started seeking access to arable land, both for cultivation and for animal husbandry, this proved much more problematic. Land resources had become scarce in the Zambezi valley; first as a result of the creation of the lake and the relocation of thousands of people, and since the late 1970s as a result of increased returns of labour migrants. From the mid 1980s traditional chiefs and headmen spent much of their time trying to settle land disputes, which increasingly also involved foreign fishermen. From the point of view of the chiefs, a more permanent settlement of the fishers would reduce the level of land conflicts and increase their control and influence among the foreigners and their prestige in the local population.

The DoF did not have the human and financial resources needed to initiate such an ambitious plan. Nor did they have sufficient legitimacy among the different groups of fishers. Nevertheless, in 1995 many of the plan's components were introduced with surprising efficiency. By the end of that year, most of the (foreign) fishermen – despite significant resistance among some of them – had been relocated into new camps (Pearce, 1994). Some of the most basic needs such as housing facilities, landing beaches and grocery shops were established in these camps, and already existing school and health facilities were made available to the newcomers (at least to some extent). Not only measures specified in the plan were taken care of; new measures identified by the fishermen during the early part of the process were also attended to. At the same time it became more and more obvious that only a fraction of what the fishing population had been promised by various authorities, would be fulfilled. The judicial foundations for the establishment of exclusive zones had not been prepared, and the promise of securing part of the fish levies to fund the co-management activities and the improvement of infrastructure did not materialize.

When investigating the forces behind the surprising efficiency of the process, it was clear that it was groups other than the DoF that had taken a main lead in the initiation of the plan that the DoF had presented in 1994. Two groups, which at first sight would seem to only have minor interests in the co-management regime, were particularly proactive. Firstly, the Kapenta operators were very proactive through their interest organization the Kapenta Fishermen's Association (KFA), but also individually through some of its most influential members. The main concern of the white Kapenta operators was to improve the control of the inshore fishermen's activities; in particular they were interested in relocating the fishermen away from a number of islands in the lake. The operators had big problems of theft of Kapenta from their fishing rigs, which they suspected the inshore fishermen of being involved in. By removing the fishers from the islands (and thereby away from the vicinity of the rigs) they thought theft could be reduced. Besides, the operators wanted to use the islands for the drying of Kapenta and some of the operators also saw an interest in developing tourist activities on them. For these purposes it was crucial to clear the islands of other resource users and as early as the first half of 1995, five of the empty islands were leased to Kapenta operators by one of the local councils.

The traditional leadership enjoyed local legitimacy while the Kapenta operators had financial resources. With the silent consent of the formal authorities it was to a large extent these two groups that gave momentum to the relocation of the inshore fishermen, which was a prerequisite for the initiation of the new management plan. The fishermen themselves remained divided in their support to it. On the one hand, it was clear that they would lose important fishing grounds, in particular by abandoning the islands. On the other hand, a lot of promises – including more regulated access to land – had been made, which were very valuable in the eyes of the "foreigners". However, when they lost access to important fishing grounds, and the promises did not materialize, many fishermen decided to leave the inshore fisheries.

To sum up, more promises had been made during the preparations than proved possible to keep; the chiefs were unable to secure improved access to arable land even if they had wished to, and the local government could not (or would not) provide the funding and the infrastructure. Slowly, some of the fishermen started to return to the islands, while others followed those who already had left the Kariba fisheries. By the end of the 1990s the remaining number was not much higher than what it had been in the early 1980s. In conclusion, one may say that it has been the competition for and regulation of access to arable land, tourist areas and to some extent the Kapenta produce, and not inshore fisheries resources, that have been the main factors explaining the reductions in fishing effort in Lake Kariba during the 1990s. (*Source: Jul-Larsen, 1995; Malasha, forthcoming.*)

In sociological and social anthropological literature on natural resources management in local communities, access-regulating mechanisms are often classified according to notions such as "community-based management systems".[12] Such general classifications can be useful, but as we will point out, they can also be problematic. One problem is to identify what kind of social institutions should be included in a system of fisheries management. The Kariba case illustrates how difficult this can be, and we are not convinced that the concerns and competition for arable land, tourist areas and theft control can be fruitfully integrated into something called a community-based management system for the fisheries. Another problem is that when we as outsiders construct something we refer to as a local management system (the population concerned do not refer to such arrangements as management systems), it may create an impression that "something exists which does not exist". As a notion, "Community-based management systems" automatically leads us to imagine the existence of structures intended to solve long-term ecological or social problems. However, whether the access-regulating mechanisms we identified in Lake Malombe and Lake Kariba can be considered as mechanisms intended to regulate fishing effort in a long-term perspective, or simply as results of people's use of institutional means in their daily struggle and competition for access to natural resources, is not easy to determine. Nevertheless, the answer is crucial if we want to know whether such arrangements can become effective parts of more long-term management institutions.

A related problem is that the notion of "community-based management" may lead to over-emphasis on the collaborative aspects of access-regulating mechanisms at the expense of a focus upon competition between individuals and groups. Elsewhere (Hviding and Jul-Larsen 1995:28-42), it has been underlined that community-based management systems always reflect people's ability and wish to establish collective action towards the exploitation of the resources, at the same time as they reflect people's inclination to compete and struggle internally for access to the same resources.[13]

Resource management as a result of collective action is well documented in the management literature (for a review, see e.g. Acheson, 1981), which often describes how the management arrangements are reflected in practices based upon various kinds of local ecological knowledge.

---

[12] Over the years, a panoply of different names and acronyms have been used to refer to such systems in the fisheries, such as "traditional fisheries management", "territorial use rights in fisheries", "sea tenure" or "customary marine tenure". We shall not enter the debate of the pros and cons of the different notions here.

[13] As such, community-based management is no different from other (e.g. state-based) management systems.

An example also emerges from this study when we emphasized the tendency to increase exploitation when the ecological productivity is high. As already mentioned in Chapter 1, this pattern of behaviour closely resembles the patterns of resource utilization observed in sectors other than fisheries. In line with the findings of Ellis and Swift (1988), which show that this "opportunistic" practice is an ecologically quite sound adaptation in pastures, Chapter 5 (examining the biological and ecological effects of increased fishing effort) shows that this is valid for fisheries in SADC freshwaters as well. People's knowledge about the ecological rationale of "opportunistic fishing", is evident in Zambian fishermen's long-lasting and collective resentment against the authorities' prohibition of certain fishing methods, such as 'Kutumpula'. The history of fisheries management (see Chapter 3) is full of examples where formal management authorities have implemented and enforced prohibitions without any form of empirical investigation. In the case of Kutumpula, more recent research (Kolding et al., 2003b) has demonstrated that the assumptions behind the banning of this fishing method were groundless, and that many fishermen actually possessed the necessary knowledge to prove this (see also Malasha personal data). When fishermen demonstrate such extensive knowledge about the ecology of their fisheries, it is perhaps not so surprising that there is often resistance in the communities to the introduction of certain types of new gear, and that they develop local regulating mechanisms to avoid introduction of gear that they consider harmful for the collective benefit of the fishery. For example, in a situation where local Zimbabwean fishermen in Lake Kariba were in conflict over most things and where it was extremely difficult to establish any kind of collaboration, they managed to effectively keep beach seines away from their shores.

Generally speaking, we have found that local access-regulating mechanisms are far more effective in preventing or limiting population-driven rather than investment-driven growth of fishing effort – particularly if the investment-driven growth comes from within the community. This is quite clear if we look at how the access-regulating mechanisms in Lake Malombe have successfully limited newcomers' investments in the fisheries, but how they have been unsuccessful in preventing the existing boat-owners increasing their investments in new and more gear. Also on the other lakes, it is the population-driven growth in fishing effort that seems particularly affected by local access-regulating mechanisms. This leads us to think that such mechanisms must often be interpreted as expressions of people's struggle to control their access to the resources, rather than more long-term arrangements aiming to conserve the resource base.

4.2.3 Capital investments in the fisheries

Finally we observe that where investment-driven intensification has taken place through the introduction of more capital-intensive fishing methods, this seems to reduce the possibility for population-driven growth in effort. In Lake Malombe the shift from gillnets to a variety of much more capital-intensive seining methods, has substantially increased the entry fees into the fishery and thereby reduced the number of potential resource users. Despite a considerable increase in effort, the number of fishermen in the 1950s and 1960s was probably higher than it is today, and during the 1980s the number of fishermen remained fairly stable (see Figure 3.5). This is a type of effort development that can be observed in most places where the fisheries have undergone major technological and organizational changes. One example of increasing differentiation between fishermen and exclusion of the poorest when the fishery becomes technologically more advanced, can be found in the difficulties of participating that Norwegian small-scale coastal fishermen faced when the fishery became more capital- and technology-intensive from the 1930s onwards (Brox, 1990). However, since investment-driven growth is still not very common in SADC freshwater fisheries, this mechanism is of little importance as a limiting factor for increased participation in the fishery at the moment. In the next two sections, we will look in greater detail into the causes behind what many scholars and policy-makers have viewed as "lack of development". As our findings show, this view needs much refinement.

4.3 Factors impeding investment-driven growth

Since much of the literature on management of African fisheries more or less assumes such a growth, the findings in Chapter 3 of very limited investment-driven growth in SADC freshwater fisheries may appear surprising. However, confronted with the huge literature on economic development in sub-Saharan Africa it can hardly be considered an astonishing conclusion. Economic growth in Africa has been very slow for many decades, particularly in the rural areas where freshwater fisheries are located. Much of the research on African economic development has been focused on identifying constraints to this phenomenon and it is within this literature we have to seek an answer to why the investment-driven growth has remained so limited, despite the increased demand for fish which is being registered in all the sub-Saharan countries.

Broadly speaking, one may say that the literature focusing on the problems in economic development contains two main currents of thought – those who put most importance on the impact of external power groups and those who focus on internal cultural or structural constraints. In his attempt to reveal the bottlenecks for vertical growth in the traditional Norwegian cod fishery, Brox refers to two theoretical models. One is the classical peasant model represented by Wolf (1966) and Shanin (1971) which mainly focuses on the power relations and the exploitation of the peasants by the wider economic systems in which they operate. The other theory is related to Chayanov's theory, from the beginning of last century, of peasant economic behaviour, which in more recent times has been taken up in James Scott's model of a "moral economy" (1976). Here the constraints are connected to cultural attitudes and values in the peasant community. Brox finds that these two explanatory models to a large extent explain why fishermen in northern Norway for a very long time did not seek to accumulate or invest in new gear. The combination of middlemen who extracted and controlled the resource rent, and an economic rationale among the producers aimed at reaching specific production targets rather than general expansion, led to a situation where there were few incentives among fishermen to invest in new gear.

Though peasant theory to some extent explains "lack" of development in northern Norwegian fisheries a hundred years ago, it seems less relevant for an explanation of the situation in SADC freshwater fisheries. Peasant theories based upon the existence of clear politico-economic hierarchies and economic exploitation seem of limited value in the study of rural Africa where these hierarchies are often multiple and combine lots of different interests. Our own empirical material supports this. We have not observed any direct economic exploitation of fishermen in the sense that the values they create were being systematically appropriated by other groups of actors, e.g. traders. It is more common that traders or middlemen have very little control over the harvesters and in most cases it seems as if the fishing population is more in control of their daily lives than the fish traders are. This is reflected in a phenomenon appearing in several of our case studies (Hara and Jul-Larsen, 2003; Overå, 2003): trade is often used as a stepping stone for accessing fisheries, while the opposite rarely happens. Although they try, traders have big problems gaining control over the production process by extending credit to fishermen. Instead of securing a stable supply of fish, the indebted fishermen tend to disappear and traders who attempt this strategy most often lose their capital.

The debate about whether people in rural Africa primarily act as 'economic agents' or whether they act on the basis of target-oriented principles of a moral economy, never seems to fade out completely. Last time the debate revived was in the 1980s when Göran Hyden introduced the concepts of "the economy of affection" and "the uncaptured peasantry" (1983) to explain why the agricultural policies in Tanzania were so unsuccessful. In our view, this debate is not particularly fruitful. Although the fact that investment-driven change (when it occurs) mostly involves "foreigners" may indicate some relevance for the notion of a local "moral economy", it does not explain why virtually all investment attempts (that actually do take place) tend to be temporary and are followed by reduction in investments and eventually a collapse of the enterprise. The Kapenta fishery in Lake Kariba, for example, illustrates the great constraints on technological and economic development in the southern Africa context, despite clear ambitions towards investments and expansion among the entrepreneurs (Turid Bøe, pers. com.; Overå, 2000). These observations indicate that the problem is far more complicated than the question of individuals' personal behaviour, attitudes and aims. In this context, it seems more fruitful to investigate in further detail the institutional landscape – i.e. not only the rules but also the underlying values and norms – under which the fisheries (including production as well as processing and distribution) operate. Therefore, let us look at some of the factors that seem to have been preconditions in the few instances when investment-driven growth actually *has* happened in the lakes we have studied.

4.3.1 Market development

First of all, the existence of an urban and/or export market for fish has been a precondition for the possibility of maximization of the resource rent through capital investments in SADC freshwater fisheries. In Malawi, for example, the population density of the Shire Highlands has been very high since early colonial times. When plantations were established, and an export market developed in Zimbabwe (then Southern Rhodesia), it was the emergence of these new markets, as well as the growing domestic urban market, that made capital investments in the fishery interesting, both for African traders and for colonial entrepreneurs (Chirwa, 1996). In the case of Lake Mweru, it was the copper mining industry in Zambia (then Northern Rhodesia) and the Democratic Republic of the Congo (then Belgian Congo) that created new markets (Gordon, 2003). Where African traders had previously traded fish to the urban areas on bicycles on a small scale, investors with capital could operate on a much larger scale because they were given the opportunity to supply the copper mines on a contract basis. It is thus not only the existence of an outlet for fish that matters; predictable profits from the market are key for investors in their decision to invest in the fisheries.

4.3.2 Infrastructure

A physical infrastructure that makes it possible to transport fish from a lake to the market is also of vital importance. Some of the lakes in this region (such as Lake Bangweulu) are located in areas that are sparsely populated and remote from the markets. The distance between the lake and the market is not always very long in kilometres, but it can be very far in terms of hours and days spent on a trading trip, punctures, petrol expenses, bad roads, unsurpassable rivers during the rainy season and other hardships. Without doubt, the remoteness of the Zambezi Valley and lack of roads leading to Lake Kariba in the 1960s, the collapse of the little infrastructure there was during the Zimbabwean war of liberation, and deteriorating maintenance of roads as the Zambian economy declined in the 1980s and 1990s, has meant that poor infrastructure has limited investments in the fisheries of Lake Kariba (Overå, 2003). An important strategy for investors in the area in the 1990s is thus to invest in roads, transport and ice plants. Since such strategies demand large amounts of capital and political mobilization, this has primarily been a concern for the Kapenta entrepreneurs and tourist operators. However, on a smaller scale, traders who attempt to establish stable supply contracts with particular fishing camps, have in some cases managed to mobilize the labour of local residents for the communal maintenance of roads in order to ensure access to the fishing camps during the rainy season (Ibid.). Another example of the importance of infrastructure, is the building of a new road to connect Lake Mweru with the expanding urban markets in the Copperbelt in the 1930s, which was a precondition for the investment boom by expatriate entrepreneurs in the fisheries (Gordon, 2003).

4.3.3 Production-distribution linkages

Investments in fisheries often happen through vertical integration of production and distribution in order to reduce the uncertainties of fish supply (for the distributor) and market access (for the producer) (Platteau and Abraham, 1987). There are three well-known strategies of linking production and distribution:

- the boat owner invests in transportation and establishes his/her distribution network in order to control more of the market chain,
- large market actors invest in fishing equipment in order to secure and increase their own supply of fish,
- the fisher/boat owner and trader enter into a contract where the former gets a steady market outlet (and often receives equipment on credit) and the latter receives fish in return (often on credit).

This last strategy is the most common when it comes to small-scale fisheries.

As we saw in Chapter 3, the expatriates who invested in fishing vessels in Lake Malawi employed the first strategy by also investing in lorries to ease their access to the markets and increase their profits. The same is the case with the Kapenta operators when they seek to establish their lines of distribution to overcome the insecurities of the present marketing system. The investments in Lake Mweru happened through the second strategy whereby expatriate businessmen invested in new fishing technology in order to increase their supply of fish for the Copperbelt markets. The Africans who invested in fishing boats in Lake Malombe, on the other hand, relied on the large numbers of small-scale traders to sell their fish. The boat owners thus got access to the market, but their relationship with the traders did not involve extensive and profitable credit-supply contracts as such (the third strategy): to invest in fishing equipment they depended on remittances from labour migrants rather than on credit from traders (Hara and Jul-Larsen, 2003). Interestingly, the cases where investors in the fishery managed to extend their control to include the whole market chain (colonial Mweru and Malawi) seem to be the result of particular privileges given to expatriate entrepreneurs by the colonial governments (e.g. allocation of land, supply contracts and fishing concessions). When they lost these privileges (as in Lake Malawi the 1950s and in Lake Kariba in the 1960s) the expatriate investors experienced the same problems as the Africans in integrating production and distribution.

The inability of producers to link themselves up to the market in order to increase their share of the profits, as well as the inability of market actors to gain control over production seems to be one of the most important direct reasons for why investment-driven growth in effort in most SADC freshwater fisheries has been so limited. The difficulty of integrating production and distribution through relations of trust and/or control is an institutional problem that seems almost insurmountable in these fisheries contrary to what happened, for instance, in the West African canoe fisheries (Overå, 1998; Chauveau and Jul-Larsen, 2000). Hence, in Lake Mweru the processing companies do not invest on the production side, and in Lake Kariba the Kapenta operators have not been able to establish a marketing system for their produce that is independent of small-scale traders and fluctuating prices. Likewise, when they invest in marketing of inshore fish, they make sure they do not waste their capital on investments on the production side (Overå, 2003).

4.3.4 Capital

Obviously capital is a major "component" in processes of investment-driven growth. As we shall discuss later, lack of capital is not primarily a practical bottleneck; it is a central institutional problem. However, in the SADC countries, the macro-economic situation has for many years been so difficult that the potential for accumulation of capital through the fish trade has been very limited. One could almost talk about an "involuted" fish market, where a growing number of small-scale fishermen produce for a growing number of small-scale traders who market the fish to increasingly poor customers[14]. The expansion in the Chisense fishery in Lake Mweru in the 1980s could be described as such a process of involution: the increase in effort in the "new" fishery was not a result of more efficient and capital-intensive technology; it was rather the result of absorption of large numbers of unemployed women and men into both production and distribution of a commodity (Chisense) that suited impoverished consumers in a declining urban economy. Therefore Gordon (2002) characterizes this as population-driven rather than investment-driven growth in fishing effort. In such a system, there is not much capital in circulation that could have been invested in modern fishing equipment: potential investors are caught in a sort of poverty trap.

Another constraint on access to capital in the SADC countries is of course that banks and other credit institutions that provide loans to small-scale producers and traders hardly exist. This is one of the reasons why investments in fisheries have depended upon input of external capital, as in the case of Lake Malombe where most of the capital came from labour migration. In this context it is interesting to note that whereas macro-economic recession often leads to population-driven growth in fishing effort, it constrains most possibilities for investment-driven growth. For example, the reduction in investments in the Lake Malombe fisheries in the 1990s is at least partly explained by the decline in labour migration to South Africa and Zimbabwe, after the crisis in the mining economy there started in the 1980s and continued throughout the 1990s.

4.3.5 The local institutional landscape

There are many practical constraints like the ones just mentioned, that limit investments in SADC freshwater fisheries. Most of them are already mentioned in numerous pieces of research and commissioned reports. However, even if factors such as infrastructure, the organization of production and distribution, and access to capital are central to whether investment-driven growth will take place or not, this only poses the questions at another level: why do these practical problems continue to constrain investments when we know that the demand for fish has increased substantially, that most individuals and groups involved seem eager to overcome them and that great amounts of foreign aid have been spent to solve them? Why does investment-

---

[14] See Geertz (1963) for a discussion of the concept of involution.

driven growth not take place in the SADC freshwaters when such growth arguably has taken place in other small-scale fisheries in the South like in West Africa and in Asia, despite many of exactly the same practical problems?

The analytical perspective on institutional development and underlying much of the co-management thinking provides some useful understandings of these issues. Jean-Philippe Platteau stands out as one of the important contributors to the school of new institutional economics. One of Platteau's concerns has been to explain why small-scale fisheries in developing countries seem to do so well despite the emergence of modern production systems, which according to conventional economic theory are far more economically effective (Platteau 1989a and b). In order to explain this, he starts by observing that the market systems in most fisheries in developing countries do not function as conventional economic theory assumes. The 'dysfunctions' are caused by market imperfections (i.e. information about availability and prices of fish does not circulate freely and is not evenly distributed among the actors), which may be observed everywhere, but particularly in the new markets in developing countries (and SADC is no exception). In order to get access to the information required, the actors in the market have to spend time and other resources. These costs are what the economists call "transaction costs". Platteau argues that a fishery based upon "modern" market institutions implies considerable transaction costs, which may often be higher than those of a fisheries based upon more "traditional" institutions[15]. In order to control labour, get access to fish, administer credits and security systems, the actors in the market will try to reduce transaction costs by making use of various types of social control mechanisms. In an imperfect market, labour and credit must be controlled by using alternative rules of recruitment, and access to fish depends on various types of interdependent control between sellers and buyers, otherwise the actors will lose and the system will collapse. "Traditional" institutions based on various types of personalized trust such as patronage, and permanent buyer-seller relations are often more cost-effective than modern institutions in terms of overcoming all these transaction costs. Platteau's analysis shows that, where the market imperfections are considerable, traditional forms of organization may simply be more profitable for the actors than the modern ones. However, the institutions have their limitations, especially in terms of the number of workers and the amount of credit that can flow through such systems. Complicated owner/worker relations thus provide limited room for accumulation and expansion, and this constrains the ability and profitability of investments. Production technology must therefore remain on a capital-extensive level.

This type of analytical approach seems particularly fruitful in explaining why a great number of industrial attempts have failed in SADC freshwaters. In both Lake Malawi and Lake Tanganyika, one of the problems of the industrial sector has been competition from artisanal small-scale fisheries. In Lake Mweru the direct cause for the disappearance of the industrial vessels was the collapse of Mpumbu, but this does not explain why the fleet was not set to target other species. In Kariba, the semi-industrial Kapenta fishery is not competing with the inshore fishery. Nevertheless, the new institutionalist approach is useful in understanding the difficulties of expansion that this fishery has due to the highly imperfect markets (Overå, 2003).

However, the same approach is less fruitful when it comes to understanding why the investment-driven growth from within the small-scale fisheries – contrary to what we observe in West

---

[15] Platteau uses the concepts of modern and traditional in the classical Weberian way. Modern institutions are based upon impersonal relations, formalized through sets of rules and laws, while traditional institutions are characterized by personalized relations, where the rules are defined according to which persons are involved.

Africa and Asia – seem to be so slow in SADC freshwaters. Part of the problem is related to the maintenance of the dichotomy between modern and traditional institutions, which makes the approach unsuited to analysing developments and changes within the traditional sector itself. Furthermore, it tends to consider traditional institutions, such as kinship relations, tribal or clan relations and patron/client relations, as phenomena that can easily be defined according to the set of rules which guide them. In the 'real' world this is not the case. Institutionalized rules for rights and responsibilities in a rural African community, and the underlying values and norms legitimizing these rules, are *not* clear and commonly shared among community members, and this complicates the analysis of the role of institutions. Studies of institutions in rural Africa in fields other than fisheries indicate that considering the question of investment-driven growth simply as a matter of "traditional" vs. "modern" institutions is far too simple. In-depth historical and ethnographic research (Berry, 1985, 1993, 2001; Peters, 1994; Fairhead and Leach, 1996; Guyer, 1997) has documented the extent to which local rules and regulations – crucial for people's access to vital resources – are unclear and ambiguous: sometimes they even contradict each other.

Our research supports these findings. Looking at how access to fish is regulated among potential buyers, we find that many different rules may be applied at the same place at the same time. In Lake Kariba a wife will normally have certain rights to the fish of a fisherman, which she will then dry and sell on the market in Lusaka. But other members in the fishing community may also claim fish from him with some sort of legitimacy. They may be relatives or they may have helped the fisherman when he established himself, or they may have drawn his nets when he was sick. In addition, a number of traders may also have rights in his catch. Some may be his relatives, another may have been particularly helpful at certain occasions like in a marriage arrangement or when one of his children got sick and needed medical attention. Others again may have helped him with small loans of goods or money, which gives them some rights in return. At the same time, no rules exist as to how much of the catch each of the entitled people can claim, so this is generally open for negotiation and discussion and will depend on a lot of factors. A fisher/fish buyer relationship is thus often far less defined and includes many more actors than many scholars – including the neo-institutionalists – tend to believe. The relationship is based upon a wide range of often conflicting rules and values, and this makes agreements and contracts very unpredictable. Although this situation may secure at least some access to fish for a large number of people, it is not well suited for those who want ensure increased and stable fish supplies. If the fisher in the example above wants to increase his production and needs credit, the traders will definitely be reluctant to give it to him: their extension of credit will not automatically guarantee their access to his fish. The traders will still have to compete with all the others who are entitled to some of the catch, and whom the fisher cannot ignore.

Another important aspect of the ambiguous institutional landscape is the relationship between owner and crew. Even if owners continuously try to build stable labour relations upon existing mechanisms of social control like kinship relations, they are often unsuccessful. The crews will often claim that they are being exploited in ways that are not acceptable. As a result, they launch activities that jeopardize the basis of their own source of employment and income. A good example is the way in which most of the crewmembers on the Kapenta rigs in Lake Kariba are probably involved in thefts of fish from the owners. They do this with reference to how bad the working conditions are. Many people in the fishing communities support their views and their actions and it is therefore limited what owners can do in order to secure their catches. Also in Lake Malombe, all owners continuously complain about their lack of control over crews in exactly the same way as Crispo Gwedela did to the "Commission of Enquiry into the Fishing

Industry" almost 50 years earlier[16]. Their testimonies are a long story of attempted thefts and destruction of fishing gear and other material (Hara and Jul-Larsen, 2003). The Malombe community's sanctions against the presence on the lakeshore of "foreign" owners when their boats landed the catch, is yet another example of the ambiguity within the local institutional landscape. Workers shifted from owner to owner, and they also developed a system whereby owners (especially those who were not indigenous to the area and those who were women) were excluded from selling the catch. Owners' lack of control made the cost of labour prohibitive. From the point of view of the owner the problems they faced qualify as theft and sabotage, but in the view of the fishermen and other community members, this was not necessarily the case. This institutional problem made it risky for investors to rely fully on fishing, and they therefore tended to diversify their investments in several sectors outside the fisheries rather than in the accumulation of fishing gear. In both of these cases, the conflicts between owners and workers represent something much deeper than resistance by individuals to accepting 'the rules of the game'. The problem is that the rules themselves are unclear and ambiguous, and this makes it possible for all parties to find legitimacy for their actions. It is not difficult to imagine how such a situation affects the owners' interest and willingness to increase their investments. Similar studies in the agricultural sector in Africa have led the anthropologist Sara Berry (1993) to conclude that institutional ambiguities and contradictions related to the access and use of labour, probably more than those related to access to land or financial resources, represent the most important constraint for economic growth in rural Africa.

Only by introducing this type of perspective can we understand why the Kariba inshore fishery, the Bangweulu swamp and the Chilwa fisheries have developed so little in technological and organizational ways. As long as the local institutional landscape does not improve so that new types of mutual trust and reliable behaviour can emerge among the actors, it is very difficult to see how investment-driven growth in these fisheries can take place. But the constructivist institutional perspective has important bearings on how resource management can be conceived. In this type of institutional landscape, the introduction of any externally driven measures of management will have few chances of success. On the contrary, one may easily foresee that such measures may lead to a further proliferation in ambiguities in the local institutional landscape. Internally driven changes may have a better chance in this respect, but as we demonstrated in the case of Kariba, it is probable that internally driven measures have their basis in the interests of specific groups and not in those of the community as a whole. We shall return to these issues in Chapter 6, but first we shall look at the consequences of increased effort on the regeneration of the fish stocks.

5. THE EFFECTS OF FISHING AND ENVIRONMENTAL VARIATION ON THE REGENERATION OF FISH STOCKS

5.1 Introduction

Typical explanations for the decline or collapse of fish stocks are that the fishery puts too much pressure on the stocks or that the environment changes, changing the productivity of the stock[17]. In the management of the African freshwater fisheries examined here, the traditional focus has been on fishing pressure with only limited considerations for environmental causes of changes in

---

[16] Record of the meeting of the commission of inquiry into the fishing industry held at the Court House, Fort Johnston, on 8 and 9 June 1956, page 17-20, MNA/COM/9/4/2.

[17] A third explanation is that the fish move somewhere else so that they cannot be captured. In the freshwater systems discussed here this will generally only be a seasonal phenomenon during fish migrations.

stocks. There is a long history of regulations against the allegedly detrimental effects of various fishing methods. Alerts about the effects of the ever-increasing fishing pressure are of a more recent date, and attempts to set sustainable levels of effort have been the goal of much research. Although the various attempts to regulate fishing effort have often proved to be fairly ineffective (Malasha, 2003) many African fisheries continue to thrive. Changes in catch composition and of target species have often taken place, but few stocks have collapsed or declined severely (e.g. *Oreochromis* spp. in Malombe; *Labeo altivelis* in Mweru; large *Lates* spp. in Lake Tanganyika), while in other cases, stocks previously thought to be endangered returned. For example, *Oreochromis mweruensis*, after very low stock levels in the 1970s is once again one of the most important components of the total catch of Lake Mweru despite an increase in effort of a factor of 5 (fishermen) to 10 (gillnets) over 30 years. The question then is why in African freshwaters do many fish stocks not collapse or yields do not decline, despite substantial increases in fishing pressure within the ranges observed in the past 50 years? How come some of these fisheries seem to be extremely resilient to increased effort while others are not and are certain fishing methods more harmful in some situations than in others?

This chapter will start by discussing the results and value of classical stock-assessments, where effort is the only variable, in the freshwater systems studied. Then, taking into account the often overlooked aspects of environmentally driven natural variability in fish stocks, we will examine the evidence on long-term effects of increased effort and discuss whether observed changes in fish stocks are a result of increased fishing pressure or are mainly environmentally driven. These observations lead to a holistic top-down approach with which the dynamics of effort development in African freshwaters can be related to both anthropogenic and environmentally driven changes in stocks. This approach entails addressing the impact of fishing on ecology and stock regeneration in three steps:

- *System variability*: the effects of environmental drivers on different scales of the fished ecosystems, indexed by changes in water level, can be used to classify systems based on fundamental ecological principles along a gradient of environmental stability from constant to pulsed systems.

- *Susceptibility of fish species to fishing*: the typical environmental variability each of these systems undergoes will be reflected in the biological responses of the fish species to it. Based on this, predictions can be made about the potential effects of fishing.

- *Selectivity and scale of operation of fishing patterns*: all fishing methods are selective and fishermen have various options to stabilize day-to-day catches. These options are related to system variability and the multispecies character of the fisheries studied. By discussing the scale of fishing operations relative to the ecosystem harvested, the development of fishing effort in different African lakes and its impact fish stock regeneration can be characterized.

5.2 Classical approaches

Rational exploitation of fish stocks involves the control of fishing mortality (effort and fishing methods) in such a way that annual catches of specific stocks can be continued indefinitely according to pre-determined objectives related to the productivity at different stock levels. The catch-effort curve of sustainable yields (Schaefer, 1954) exemplifies this approach: at any level of fishing effort up to the level where the 'surplus yield' is maximized, a yield can be found that is theoretically sustainable and stable (Appendix section1). Which level of fishing effort is chosen depends on a number of strategic objectives (Salz, 1986) such as securing a minimum biomass, maximizing food production (MSY), maximize the resource rent (maximum economic yield, MEY) or employment. Of these objectives, the concept of maximum sustainable yield (MSY) at which effort levels should be set in order to maximize food production has gained most prominence. Various models estimating MSY, or maximum yield per recruit, have been used extensively in African freshwater fisheries, and the concept of MSY has formed part of the research goals in many fisheries development projects as well (Kolding, 1994).

It is important to reiterate the biological assumptions of the surplus-production model most often used in the African context, the Schaefer model, because of the many policy implications that have arisen from these assumptions:

- Density-dependent logistic growth under environmentally stable conditions, which means constant carrying capacity (see Appendix 1).
- It is a single species model, with man the only predator.
- In the traditional application, it presupposes a continuous steady state, which means instantaneous equilibrium between catches and surplus regeneration.

These biological assumptions have been questioned for a long time in an extensive literature (e.g. Larkin, 1977; Sissenwine, 1978; Kolding, 1994), and are questioned again by the various case studies in this report. For now we will just note that natural variability in fish stock levels due to environmental variation results in changing values of the underlying biological parameters of the model: the intrinsic growth rate (r) and the carrying capacity (K) (see Appendix section 1). This will result in considerable uncertainty around the estimated sustainable yield curve, as constant conditions are not present and surplus production is not regulated by effort alone. Systems that experience less environmentally induced interannual variability may conform better to the underlying steady-state assumptions of the model, possibly resulting in more informative results from stock assessments. However, in systems with a large environmental variability, attempts to relate trends in fish stock levels to fishing alone may be more difficult. Such trends will be hidden in environmental variation or, in statistical terms, noise or error, resulting in what is often called 'process error' (Caddy and Mahon, 1995), that is variability caused by the unknown states of nature. Moreover, the multispecies and multi-gear situations encountered in most tropical small-scale fisheries, in particular in freshwaters, make the application of standard models even more problematic.

In all the lakes of our case studies, except Lake Mweru, previous attempts have been made to estimate MSY in order to set the level of effort at which the fishery would operate at maximum efficiency, or other sustainability levels (Mwakiyongo and Weyl, 2001) sometimes by gear type. Both Malombe and Kariba are highly instructive examples with regard to the limited informational value that such assessments have (Boxes 5.1 and 5.2). In Malombe, MSY from surplus production models has been estimated for the Chambo (*Oreochromis* spp.) and Kambuzi (haplochromine cichlids). The application of models was successful in the sense that they conveyed the message of limits to the system overshot severely. However, additional information was needed to explain the particular time sequence of catch and effort levels, as their behaviour did not conform to standard model predictions.

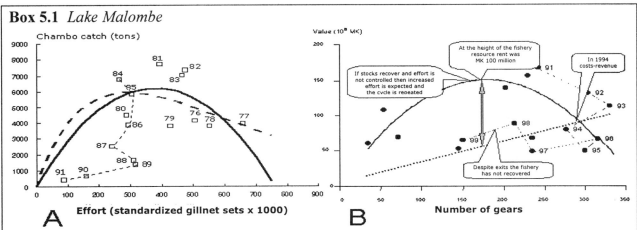

Box 5.1 *Lake Malombe*

Figure 5.1 *Surplus production models applied on the Lake Malombe fishery. A. Schaefer (thick continuous line) and Fox (thick broken line) models of the gillnet fishery on Oreochromis spp. (Chambo) of Lake Malombe (modified after FAO, 1993). The curves are fitted for the period 1976 to 1985. After 1985 the fit does not hold, as the Oreochromis stocks collapse due to the effect of the Kambuzi seine and Nkacha net fisheries. B. Gordon-Schaefer analysis of the purse seine (Nkacha net) fishery on Haplochromis spp. (Kambuzi) of Lake Malombe (modified after Weyl, 2001).*

Box 5.1 *Lake Malombe* (continued)

The stocks of Chambo (*Oreochromis* spp.) collapsed after 1984 (Figure 5.1A), while in subsequent years the haplochromine stocks (Figure 5.1B) seemed to undergo the same fate. A reduction in the level of effort took place irrespective of government regulations and did not lead to a recovery of stocks despite what was predicted. In the case of Chambo, the standardized gillnet effort levels decreased by a factor 7, in the case of Kambuzi, the purse seine effort decreased by a factor 2. This lack of recovery is totally against the underlying assumption of balance between yield and effort, which renders the use of surplus production models hardly appropriate. In the case of Chambo, the cause of the decline was explained *ad hoc* to be heavy fishing by other gear on juveniles (technical interactions) and possibly the destruction of juvenile habitat. Five management scenarios were developed with predictions on the duration of the recovery of the Chambo stocks (FAO, 1993), basically entailing different levels of effort for the various gears employed. None of these happened, but large changes in gear use and effort took place, including a slight change in mesh size of the small purse-seine nets after protracted negotiations, without recovery. The continuously decreasing water level over the period the models covered was not taken into account. In the case of Kambuzi, no possible causes have been put forward, but recent observations indicate that the decreasing water levels over the period of stock decline may have played a role (O.L.F. Weyl, pers.com.).

In both cases, the interpretation of the results has been done, *without* using the model. In other words one could have arrived at the same time analysis without drawing the curves and by just plotting yield against efforts.

Though an idea on the limits of the system is given, the models only gave them when fishermen had already reached them. Thus, despite the fact that the lake:

- is small,
- it has had relatively low environmental variability over the period of data used for the Chambo model;
- it has had a spectacular development in fishing effort;
- it has a solid information base compared with many other lakes in the region; and
- it has a relatively 'simple' fish community,
- the models have not fared very well and much of the relevant information on the functioning of the system could have been obtained without them. Additional effects such as those of the main fishing gears on juvenile *Oreochromis* habitat remained out of the main discussions. No updating and regular evaluations of the validity of the model predictions from the stock-assessment models have been made. This indicates that a learning effect on the behaviour of the Malombe system, based on further refining the information obtained from the models has been minimal.

Box 5.2 *Lake Kariba*

The Kapenta fishery on the Zimbabwean side of Lake Kariba experienced a decline in catch per unit effort during the late 1970s (Karenge and Kolding, 1995a), although the long-term trend is not significant. Fear of overfishing or indications of fully exploited resources have repeatedly been expressed in Lake Kariba (Marshall, 1981; Kenmuir, 1982; Marshall, 1985; Machena and Mabaye, 1987; Marshall and Langermann, 1988; Moyo, 1990), and it was commonly believed that the observed decline was due to overexploitation. However, these opinions have been contested (NORAD, 1985; Ramberg *et al.*, 1987; Marshall, 1992; Kolding, 1994). Though decreased catch per unit effort and changes in species compositions sustained it, the notion of overexploitation appears to be based on inconsistent and misleading MSY calculations.

Several of the attempts at calculating potential yields were based on the simple empirical equations of Gulland (1971), Henderson and Welcomme (1974) and Melack (1976). When comparing the result of these estimators, or when matching the predictions with the observed yield, large discrepancies are seen (Figure 5.2). Usually such discrepancies are explained by inadequacies in the data and/or various *ad hoc* hypotheses[18]. But the empirical equations are constructed from observed yield in various African lakes, irrespective of the fishing intensity level. In the early 1970s, when the correlations were made, yields may have been at their sustainable maximum, but judged upon later developments in most systems they most likely were not. Thus, the classic empirical equations only give an estimate of the observed mean yield in various systems in comparison with environmental parameters at the time of investigation. Theoretically, however, they say little about how far a fishery can expand in terms of sustainable yields.

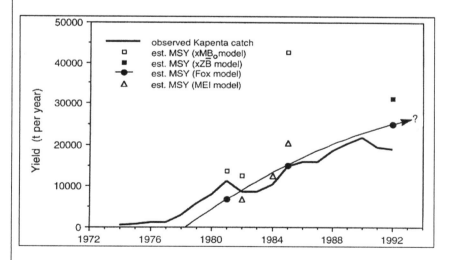

Figure 5.2 *Observed Kapenta yield on the Zimbabwean side of Lake Kariba plotted against 4 groups of various MSY estimates.*
MSY calculated from Gulland's (1971) formula (xMB_o, where $x=0.5$.
MSY calculated from the xBZ model (Cadima) with $x=0.5$, $Z=5$ yr-1 and $B=25,000$ t. (the MSY values from (1) and (2) are halved to include the Zimbabwean area of the lake alone).
MSY calculated from Fox's (1970) surplus production model.
MSY calculated from Henderson and Welcomme's (1974) MEI relationship.
Sources: Marshall (1981, 1984, 1985, 1992), Marshall et al. (1982). Reproduced from Kolding (1994).

---

[18] A typical example of discrepancies are observed in Lake Tanganyika where the pelagic yields are many times the predicted value (see Coulter, 1991).

Three applications of the traditional surplus production models (Schaefer, 1954; Fox, 1970) during the past 20 years in Lake Kariba (Figure 5.2) show that estimates of the MSY of the pelagic Kapenta (*Limnothrissa miodon*) have increased along with the increase in catches. Curiously, each past estimate was very close to the actual maximum catch over the observed period and suggests that the current level of effort, at any time, is around its maximum (Marshall, 1992). While each investigation concluded that the fishery was now near its potential maximum, later increased effort and catches also turned out to be sustainable. Marshall *et al.* (1982) and Moyo (1990) also applied surplus production models on the inshore fisheries in Lake Kariba. They arrived at the same peculiar result that the predicted MSY was very close to the actual current mean catch and they concluded that the effort level was now at its limit. The problem however, is not peculiar to the Kariba fishery, but a general feature when fitting surplus production models under equilibrium assumptions to fisheries with only data points on the ascending side of the yield-effort curve. Hilborn and Walters (1992) describe a similar situation and conclude that it is simply not possible to find the top of a yield curve without going beyond the top: "You cannot determine the potential yield from a fish stock without overexploiting it". This, in fact, is illustrated by both the Malombe and Kariba cases[19]. Furthermore, the latest attempt to apply the Schaefer surplus production model (Anon., 1992), revealed that effort did not explain the variation in catch rates in the offshore Kapenta fishery. Consequently it was found to be impossible to apply a surplus production model with any confidence on such a changeable stock.

---

[19] The problem is based on the difficulties in estimating the intrinsic growth rate (r_m) in the underlying model from catch and effort data, since the true value of r_m is only defined when the stock levels are extremely low (see eqn. 3 in Appendix 1, Box 1), in which case they would be seriously overfished.

Table 5.1 *The relative effect of hydrological changes and effort on catch rates in three systems studied The statistical regression model used is: Annual mean catch rate$_{ijk}$ = overall mean + effort$_i$ + lag(hydrological variable$_j$) + effort$_i$ x lag(hydrological variable$_j$) + residual variation$_{ijk}$. Only significant effects are retained in the model and shown here as positive (+) or negative (–) effects. In some instances the hydrological variables and effort were confounded meaning that both parameters were significant alone, but in the total model one or the other became non-significant depending on the order they were entered into the model. In such situations it is not possible to quantify the relative effect of both parameters simultaneously. N = number of years. See text and Zwieten and Njaya 2002, 2002a and Kolding et al., 2003 for further explanation of the model.*

Dependent variable			*Independent variables*						
Annual mean catch rate (CpUE)			**Water level**			**Effort**		**Inter-action**	
System Variable	N	Lag	Variable	Sign	%	Variable	Sign	%	Sign	%
Lake Kariba Zimbabwe										
Artisanal (kg/net)	27	1	Annual change (Δ)	+	29	# fishers	–	44		
Artisanal (kg/net)	27	1	Annual change (Δ)	+	19	# nets	–	26		
Experimental (kg/net)	29	0	Amplitude	+	39	# fishers				
Experimental (kg/net)	29	0	Amplitude	+	39	# nets				
Kapenta (ton/night)	26	0	Mean	Confounded		# boats/night	Confounded		–	56
Lake Kariba Zambia										
Artisanal (kg/net)	20	0	Mean	+	32	# fishers				
Artisanal (kg/net)	20	0	Mean	+	32	# nets	–	9	–	8
Experimental (kg/net)	20	0	Mean	+	31	# fishers				
Experimental (kg/net)	20	0	Mean	+	34	# nets	–	23	–	11
Kapenta (ton/night)	18	0	Mean	Confounded		# boats/night	Confounded		–	60
Lake Malombe Malawi										
Kambuzi seine (kg/haul)	18	3	Minimum	+	47	# nets	–	32		
Gillnet (total kg/100m net)	18	0	Maximum	+	68	# 100m nets				
Gillnet (kg Oreochromis/100m)	18	3	Minimum	+	77	# 100m nets				
Lake Chilwa Malawi										
Gillnet (kg/100m net)	18	1	Mean			# 100m net	–	38	+	38
Gillnet (kg Oreochromis/100m)	18	0	Maximum	+	20	# 100m nets	–	26	+	30
Seine (kg/haul)	18	0	Maximum	+	16	# hauls	–	54		
Longline (kg/hook)	18	0	Maximum	+	9	# hooks	–	57	+	20
Trap (kg/trap)	18	0	Minimum	+	17	# traps	–	43	+	13

Box 5.1 illustrates how the systems have evolved under increasing fishing pressure, and that, where stocks have collapsed, MSY in fact was found empirically. In Lake Kariba, the various attempts of estimating MSY and corresponding effort levels have all been unsuccessful in retrospect, partly because of model limitations and partly because of data limitations (Kolding 1994, Box 5.2). With hindsight, it can be maintained that several misconceptions and mistakes have been made in the past when evaluating and explaining the observed variations in the yields from Lake Kariba. These errors, however, are partly understandable when working under the conventional 'stability' paradigm, and using the established yield models.

Both cases represent situations where a good biological information base is present. In other cases (e.g. Lake Chilwa) some of the monitored information, such as fishing effort, was considered so unreliable that attempts to fit simple assessment models were regarded as completely unsatisfactory (Tweddle, 1995). Clearly, under such circumstances, and additionally in fluctuating systems, the very concept of 'MSY' as a management objective becomes highly questionable (Kolding, 1994, 1995). Still, the MSY concept remains deeply rooted in planners, administrators and research personnel, but so far it has mainly created misunderstandings or false expectations. Attempts to determine MSY in environments with a large, but to some extent predictable variability in productivity levels have failed. On the contrary, it can be generally said that yield (or Yield Per Recruit), models have not been of much use in any of the freshwater fisheries in Africa (Coulter, pers. com.).

How large the effect of environmental variability on stock levels in fact is can be seen in Table 5.1. In this table we have compiled the results of the relative effect of hydrological changes and fishing effort on catch rates in three studied systems. In all cases the analysis was performed in two steps (see Zwieten and Njaya, 2003; Zwieten et al., 2003a and Kolding et al., 2003 for further explanation). It takes time for any species to 'grow' into a gear before it can be caught (see below). This means that an effect of higher productivity on stock levels will only be observed after some time – usually between zero and four years depending on the size of the species caught in the fishery. The effect of hydrology on such 'recruitment variation' is the lag phase between annual hydrological variables and catch rates. If there is no time lag this means that effects of hydrological changes is felt by the fishery within the same year. In the second step the statistical model (stepwise multiple regression) is performed according to the lag phase found. The relative effect of effort or water level on catch rates is examined through the proportion (%) of the total variation explained by each independent variable in the model. It is clear from Table 5.1 that in nearly all cases hydrological changes had a highly significant positive effect on annual mean catch rates, meaning that catch rates decrease or increase according to similar changes in water levels. Increased fishing effort has in many cases the expected negative effect on catch rates. However, the table makes clear that such an effect only explains part of the total variation in catch rates, and in any case this part will often be difficult to distinguish directly from the effects of the environment on stock levels.

5.3 System variability: water level as environmental driver

In the long history of research on fisheries it has been debated whether fishery collapses are due to overfishing or due to environmental change (e.g. Corten, 2001). Large changes in productivity of fish resources may be due to environmentally driven processes, in particular where large changes of nutrient input occur. The most conspicuous external drivers of nutrient inputs in all of the systems in this study are long-term, interannual and seasonal fluctuations in water level and river inflow. Biological productivity, from algae to fish, depends essentially on the nutrients introduced by the annual flood regime, including the nutrient mobilization through flooding of lake margins or associated floodplains and swamps. Such systems are also called allotrophic riverine systems (Kolding, 1994). The water budget, monitored through water levels, can thus be used as a proxy for abiotically driven changes in productivity (Kolding, 1992; Karenge and Kolding, 1995a). Most lakes have rivers flowing into them, or are at least affected by rainfall runoff from land. It thus depends on the ratio of the inflow to the size (volume) of a system – the flushing time – how much a system is externally driven by water and nutrient inflow. For example, the average flushing time of Kariba is 2.6 years and of Mweru approximately 3 to 4 years. In contrast the flushing times of the large lakes Tanganyika, Malawi, Victoria and

Turkana are respectively 7 000, 750, 140 and 12.5 years (see Table 5.2). In these large systems (except Turkana) biological productivity is predominantly determined by an internal supply of nutrients through vertical mixing and nutrients from the atmosphere. External drivers in the very large systems refer to wind stress (e.g. Spigel and Coulter, 1996; Plisnier et al., 1999; Sarvala et al., 1999) and seasonal changes in solar radiation (Talling and Lemoalle, 1998) that affect mixing processes. The much smaller lakes of our study are fully mixed, at least for most of the year, while the large lakes (except Turkana) are all permanently or seasonally stratified (that is with different layers of water that do not mix).

The magnitude of changes in productivity, in other words the relative stability, thus depends on the size of the lake relative to the inflow rate. This in its turn is a function of the catchment area, which determines how much a system depends on local patterns of rainfall. But stability also depends on the temporal scale (periodicity) of the fluctuations relative to the life history characteristics of fish species and communities affected by them. Long-term climatic changes over large areas are reflected in long-term fluctuations in lake-levels (Nicholson, 1996; Nicholson and Yin, 1998). Such long-term changes may be associated with large-scale abiotic and biotic changes, and determine the range over which a system has oscillated historically and could be expected to do so in the future. Knowledge about longer-term changes that took place in the past, even the recent past (100 years or less) is important. It gives indications about the long-term dynamics of a system that is not human-induced and it can give clues about the persistence of trends and states over longer time spans. In statistical terms, this part of the total variance in time series of catches and catch rates would be called 'red noise' in analogy with the low periodic cycles in spectral analysis.

However, in a fisheries management context the important scale is the interannual variability over much shorter timescales, in general from one year to a decade. Interannual variability gives information about the relative stability (and hence predictability) of the productivity of a lake and with that the persistence of stock levels of longer lived species in particular. This is the timescale over which year class variability of longer-lived species as a result of interannual variability becomes important. In a tropical environment, the period over which cohorts of longer-lived species are under exploitation is about three to four years, which will be seen in time series of catch rates as medium-term fluctuations or 'blue noise'. Lastly, intra-annual variability, or seasonality in water levels gives the extent of the variability in nutrient pulses that are generated each year. Stock levels of short-lived species such as the pelagic clupeids of Mweru and Kariba (but also those of Lake Tanganyika (Zwieten et al., 2002c)) and the small barbs in Lake Chilwa (Furse et al., 1979) are particularly driven by bottom-up processes resulting from seasonal variability in nutrient pulses [20].

Thus, the effects of lake level fluctuations on productivity will be operating on a range of timescales. The biomass of fish species or a of community could be expected to co-vary over similar scales. Long-term historical fluctuations in water level due to climatic changes can give a measure of relative stability over that time. Over shorter time spans, externally driven conditions may still be highly variable, which will characterize changes in the fish communities and fisheries operating on them. We thus need a measure of variability of a particular water body

---

[20] This is apparently not the case in Malawi, where top-down processes through predation are thought to dampen effects of existing seasonality in lower trophic levels of the pelagic system (Allison et al., 1995), though this conclusion is based on rather short time series.

over different timescales. In addition this measure should include the scale of the system, as it will be clear that a similar range of lake level fluctuations in large systems will have a smaller or a much more localized effect, compared to small systems. Therefore we have devised an index of Relative Lake Level Fluctuations (RLLF), defined as:

$$RLLF = \frac{mean\ lake\ level\ amplitude}{mean\ depth} \cdot 100$$

By calculating this index both for the average interannual (change in annual mean = RLLF-a) water level and the average seasonal pulse (RLLF-s), we have an index both for the average inter-annual stability of a system and the average strength of the seasonal pulse with which we can scale different systems. With each system, important measures would then be: (1) the persistence of conditions on a decade scale, and (2) the extent and duration of extreme high or low levels. Persistence is important to predict short-term trends while peak levels are expected to be important to predict year class strength. Here, we will only discuss these by visual examination of the time series of water levels. A third measure of importance would be the change in surface area of the lake as a result of changing water levels. We do not have such information, but it would be of considerable importance for lakes with low sloping shorelines as for instance in large parts of Kariba, for the smaller lakes, but also for local conditions in larger lakes (e.g. Fergusson's Gulf in Lake Turkana (Kolding, 1993a)). These measures can be used to construct empirical relations with time series of catches or catch rates.

Historical to long-term fluctuations in lake-levels could indicate large-scale ecosystem changes and resets. Large fluctuations in lake-levels have taken place over long periods of time in the Great Lakes of Africa (Table 5.2 and Figure 5.3). These lakes act as integrators over long-term changes in rainfall patterns, and in a feedback loop partly influence rainfall patterns as well (Nicholson and Yin, 2001). Historical long-term fluctuations of the large lakes are remarkably correlated over the past 200 years[21]. Smaller lakes follow the same pattern of climatic change and some of these may even dry up completely during extended periods of low rainfall, e.g. Lakes Chilwa, Malombe, Mweru Wa'Ntipa, Nakuru and Stephanie. Such droughts mean complete resets for the ecosystems of these lakes. Species diversity is related to the variability and the size of the system (Figure 5.4). For somewhat larger systems like Mweru and Kariba, climatic changes could also mean long-term changes in relative species composition, as can already be observed on a shorter timescale in large shallow pulsed lakes, for instance Lake Chad (see citations in Lévêque, 1995 or Lévêque, 1997). However, time series of catch rates of Mweru and Kariba are too short to be able to confirm such species successions. Those found in Lake Kariba appear to be influenced by lake-level changes (Kolding and Songore, *in prep*). Still, in this case succession may be largely a function of the age of the system and not of historical periodicity in changes in inflow of the Zambezi River.

Long-term changes would result in long-term fluctuations of stock levels ('red noise'). Time series of catch rates or of other indices of fish stock size are generally too short – even in temperate zones often not longer than 60 to 80 years – to show the presence of red noise. On the other hand, if over large areas and systems, similar patterns exist, then the possible existence of long-term variation can be pointed out. For instance, in the later half of the previous century, in

---

[21] Comparisons between Lake Malawi and Lake Victoria over the past 1 200 years generally show opposite trends for both lakes consistent with the most typical patterns of rainfall anomalies, that show strong opposition between equatorial and southern Africa during most years (Nicholson, 1998a).

particular during the 1970s, all lakes and river systems throughout East and Central Africa had exceptionally high water levels (Nicholson, 1999; Laraque *et al.*, 2001), in contrast to West Africa and the Sahel (e.g. Le Barbe and Lebel, 1997). Though productivity changes in the Great Lakes will, as discussed below, largely reflect changes in internal dynamics, the rise in water levels did reflect a change in climatic conditions. It coincided for instance with a remarkable rise in mean catch rates and in a more pronounced seasonality of clupeids in Lake Tanganyika (Zwieten, 2002c). Catches of the whole SADC area, dominated by the output of the Great Lakes, show a fluctuation around the main trend, that is significantly positively correlated with lake levels of Tanganyika and Malawi with a time lag of six to eight years (Figure 5.3). Such a lag is too long to be explained merely by fluctuations in stock levels, as these would have to be visible within one to four years, more so now that small pelagic species are dominating catches in most of these lakes. However, this lagged persistence in catch levels related to lake-levels could indicate a time lag in relative effort levels as well. Good catch conditions could attract fishermen over a period longer than these conditions exist and *vice versa* possibly resulting in fluctuations in total output. Such results highlight the need for good information on trends and fluctuations in numbers of people fishing to be able to understand their significance. If our inference is correct, it could mean that catch rates are in themselves a regulating factor of population-driven levels of effort.

Interannual lake levels indicate medium-term trends and persistence of conditions directly affecting the variation and persistence of fish stocks. In this shorter time perspective, local rainfall conditions in the catchment area dominate (Nicholson, 1998b) and interannual changes in water levels in Lakes Mweru, Bangweulu, Kariba and Chilwa reflect these. In the lakes of our study, the range of annual mean levels over the period examined is between 0.8 and 3.7 m, but mean change in water level (Δ WL) and variability (CV = Coefficient of Variation = standard deviation*100/mean) in water level is very different. Lake Kariba, Lake Mweru and Lake Chilwa have the largest mean Δ WL (1.30, 0.53 and 0.58 m) and CV (135%, 93% and 94%) of all the lakes examined. But the effect on the system differs considerably as is reflected in the RLLF-a: 4.3% (Kariba), 7.2% (Mweru) and 17.8% (Chilwa). Malombe and Bangweulu have the lowest mean ΔWL (respectively, 0.32 and 0.26) but Bangweulu is a much more variable system (CV=78% compared to Malombe CV= 56%). Lake Malombe can be considered as a satellite lake of Lake Malawi and largely follows its changes in water levels. Compared to other lakes in this study, it is much more stable in terms of mean annual change, variability and RLLF-a.[22]

The next element to consider is the level of persistence of environmental conditions. Examining short-term trends in lake-levels is one way to do this. A cursory examination of the time series of relative water levels and river inflow in Figure 5.3 indicates that all of the lakes examined exhibit upward and downward trends and stable periods generally not longer than five to six years, followed by peaks in water levels. As noted already, persistence is reflected in catch rates in different ways for different species depending on their longevity (Zwieten and Njaya, 2003; Zwieten *et al.*, 2003a paragraph 5.5) and the peaks are important for the year-class strength of longer-lived species. These short-term trends in environmental behaviour of the system will be reflected in short-term trends in catch rates ('blue noise').

---

[22] Note that mean depth can change considerably in these small lakes, which will affect the RLLF: in the case of Malombe we have chosen to take a mean depth averaged between periods of low and high water level. It will also affect the volume of the lake: with the two published average depth figures of 4m and 7m, the total volume of Malombe differs by a factor of almost 2!

Table 5.2 *Lake size and lake level variability ordered according to increasing RLLF-a. (See text for further explanation.)*

Lake	Size (at present flood levels)				Variability in lake levels in m					Relative Lake Level Fluctuation	
	Surface (km²)	Volume (km³)	Catchment (km²)	Mean Depth (m)	Historical			-Interannual amplitude (a) -Seasonal amplitude (s)			
					Year	Range	Description of long-term variability in water levels	Range	Mean ΔWL	RLLF (%)	
Tanganyika	32 600	18 800	249 000	580	1770 2000	36	Low water levels, app. 15 m below present, during the late 18th century until around 1840-50. Gradual rise till around 1870. Fast rise to 20 m above present around 1880 with a subsequent drop to 10 m below present level around 1895. Long-term fluctuations around present levels since then.	3.2 1.0	0.22 0.78	0.04 (a) 0.14 (s)	
Malawi	28 800	8 400	97 700	290	1800 2000	14	Water levels around 900, 1300, 1650 app. 6 m higher than between these periods. Extreme lows by the end of the 18th century, rising to present water levels (2 m lower than maximum) around 1880. Drop of app. 2 m during the start of the 20th century gradually rising to present. Rainfall patterns in the northern part of the catchment mainly determine levels	5.4 1.4	0.28 0.97	0.1 (a) 0.3 (s)	
Victoria	68 800	2 760	149 000	40	1780 2000	4	Decrease during the end of the 17th, start of the 18th century. Extreme lows around 1830. Gradual increase until app. 1860, thereafter a fast increase (with variation) until 1.5 m above present water levels around 1880. Drop immediately afterwards to early 20th century levels. In 1960 levels rose suddenly with around 2 m to the present high water levels.	2.6 2.4	0.22 0.44	0.6 (a) 1.1 (s)	
Turkana	7 560	237	131 000	31	1888 1989	20	Rise to about 15 m above present levels peaking around 1896. General slow decline during the first half of the 20th century. Later fluctuating with very low levels in 1945, 1955 and 1988.	2.0 -	0.61 -	2.0 (a) -	
Kariba	5 400	160	664 000	30	1963 2000	10	Dam across the Zambezi completed in 1958. Filling till peak level in 1963, followed by a period of large fluctuations around a mean level until 1974. From 1975 to 1981 mean levels rose with 2 m, followed by a decrease of 7 m over only 3 years with subsequent low levels. Started refilling in 1997 reaching full dam levels in 1999.	3.5 5.9	1.30 2.90	4.3 (a) 9.7 (s)	

Lake	Size (at present flood levels)			Variability in lake levels in m						
	Surface (km^2)	Volume (km^3)	Catchment (km^2)	Mean Depth (m)	Historical		Description of long-term variability in water levels	-Interannual amplitude (a) -Seasonal amplitude (s)	Mean ΔWL	Relative Lake Level Fluctuation RLLF (%)
					Year	Range		Range		
Malombe	450	2.5		5.5	1915 2000		Between 1915 and 1924 the lake dried up entirely. Between 1924 and 1934 no lake until the sand bar blocking the Shire river swept away. Mean depth 4m (Van den Bossche and Bernascek, 1990); 7m (FAO, 1993).	3.1 1.4	0.32 1.12	6.0 (a) 20.4 (s)
Mweru Lake Swamp/ Floodplain	4 650 1 500 900	38 - -	207 774	8	1956 2000	-	Follows Lake Malawi patterns: dependent on rainfall in the Chambeshi and Luapula catchment area, i.e. predominantly northern Zambia.	3.3 3.3	0.58 2.05	7.2 (a) 25.7 (s)
Bangweulu Lake Swamp/ Floodplain	2 733 5 180 12 271	9.9 - -	99 502	3.5	1956 1995	3.5	Part of the Chambeshi and Luapula River system that connect with Lake Mweru. Lake levels lowest in November/December. High water level is reached at the end of the rains, usually in April. Mean seasonal water level variation is 1.2 m. Extreme water level variation up to 2.3 m	0.8 2.3	0.26 1.20	7.4 (a) 34.3 (s)
Chilwa Lake (1972) Swamp Floodplain	680 580 580	2 - -	8 780	3	1600 2000	12	Extreme water levels during the 17th century, 12 m above bottom level, with a drop and stabilization with regular prolonged dry periods during the 18th, 19th and the start of the 20th century. Lake levels mainly determined by rainfall patterns in the Mozambican part of the catchment area.	3.7 1.7	0.53 1.19	17.8 (a) 39.7 (s)

Based on Kalk et al., 1979; Crul, 1993; Evans (1978); FAO, 1993; Kolding, 1989, 1994; Bos, 1995; Crul, 1995a, 1995b; Nicholson, 1998a, 1998b; Nicholson and Yin, 1998; Nicholson, 1999; Laraque et al., 2001; Nicholson and Yin, 2001; van den Bossche and Bernascek, 1990; Verheust and Johnson, 1998; Department of Hydrology, Zambia.

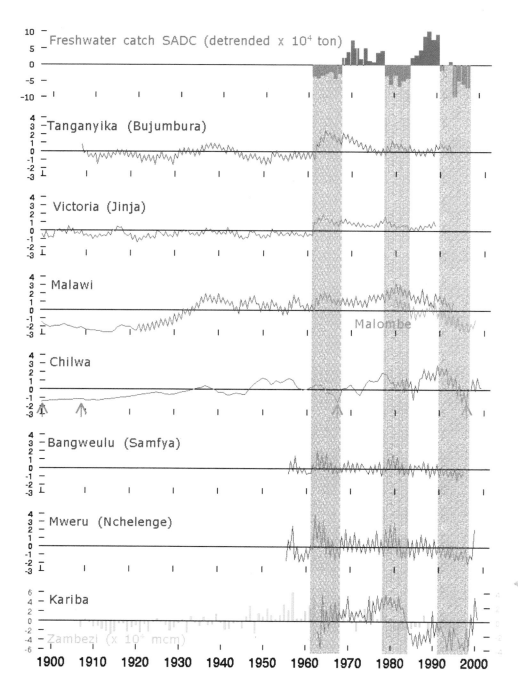

Figure 5.3 *Relative water levels of lakes Tanganyika, Victoria and Malawi and the five lakes of this study. In blue are deviations from the long-term mean of annual mean levels over the period for which data were available. Green bars are the deviations of the 89 year mean annual total inflow of the Zambezi at Victoria falls (mcm = million cubic meter)(Langenhove et al., 1998). Arrows indicate the years Lake Chilwa was reported to be dry. Malombe (orange) is hydrologically considered a satellite of Malawi: when this lake has low water levels Malombe completely dries up. The relative levels of Malombe and Malawi is shifted to be able to show their respective lake level developments. Lake levels are based on Crul, 1995a (Victoria and Tanganyika), Nicholson, 1998b (Chilwa), the Department of Hydrology in Malawi (Chilwa and Malombe), the Department of Hydrology of Zambia and own measurements (Bos, 1995 and van Zwieten (unpublished) (Mweru) and Kolding, 1994, Songore, 2001 (Kariba)). The top panel shows the variability around the trend of the total catches of the SADC region (see also* Figure 2.1*), with grey bars comparing periods of low catches relative to the trend.*

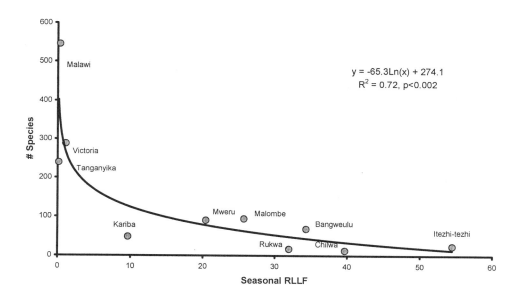

Figure 5.4 *Number of fish species in ten SADC lakes related to the seasonal Relative Lake Level Fluctuations (RLLF-s).*

Lastly, intra-annual lake levels represent seasonality in productivity. The actual (variation in) size of nutrient pulses is conditioned by the seasonal variability in water inflow and levels. Variation in size and duration of floods is important in regulating the fish productivity in a cascade of effects through various trophic levels in an ecosystem ('wave of productivity' – see Pope *et al.*, 1994), where interactions between species may dampen such effects (see below). Mean seasonal change in lake levels is highest in Lake Mweru (2.1 m) with considerable variability (CV = 38%), indicating that the size of the seasonal pulse varies considerably between years. By comparison, Lake Malombe exhibits a lower seasonal variability (CV = 23%), again indicating that also on this temporal scale the lake is more stable than other lakes in our study (Kariba CV = 49%; Bangweulu CV = 35%; Chilwa CV = 36%). Taking into account the scale of the lake, the effect is highest in Chilwa (RLLF-s = 39.7%), followed by Bangweulu (34.3%), Mweru (25.7%), Malombe (20.4%) and Kariba (10.1%) (see Table 5.2).

The potential impact of changing water levels is thus relative to the scale and size of the system:

- From long-term trends and fluctuations and complete resetting of systems acting on the composition of whole fish communities ('red noise'),
- via short-term trends and interannual fluctuations operating on fish species within communities on a short term depending on lifespan and response of species to changes in externally driven lake productivity ('blue noise'),
- to variations in seasonal pulses acting on the recruitment of species with short lifespans, triggering fish migrations and, being the starting point of the annual productivity pulse, the year-class strength of longer-lived fish.

By examining different scales of variability, we can thus position a system within a general classification of lakes in terms of system stability. Within the class of systems externally driven by water inflow, or pulsed systems, we may also have an important general indicator for changes in stocks, easily measurable, available and easily communicable. The results of our study on the effect of water levels on catch rates (where both time series are available) are summarized in

Table 5.1, which shows that trends and variability in annual mean or change in water level is dominant in explaining catch rates in all of these. It is also clear that effects of changes in water level are observed in catch rates of individual species for periods of between one to four years. In the next paragraph we consider how such variability affects individual species and fish communities on different timescales in order to explain these results. It is important to know to what degree fish yield is determined by abiotic factors (such as environmentally driven productivity changes) and by biotic factors such as predation, competition and life history characteristics of individual species.

5.4 Susceptibility of fish stocks and species to fishing under environmental variation

We will now draw up a framework to help explain why fish stocks in pulsed systems appear to be more resilient under increasing fishing effort compared to those of more constant systems. Fish stocks, like all natural populations, fluctuate in abundance with or without exploitation. A full understanding of the causes of the variations is still lacking (Rothschild, 1986), but Regier (1977) developed a conceptual framework for classifying the dynamics of fish stocks according to a number of biological and ecological research traditions. The framework defined four broad classes based on two variables along which a resource could be positioned:

- size of the resource (small to large) and
- temporal and/or spatial variability of a system (low to high).

Regier's four classes were:

- Large stocks generating large amounts of revenue in relatively stable environments. The traditional stock assessment models have largely been developed for this type of problems. However, it has long been recognized that the usefulness of advice resulting from this approach is reduced with systems that are changing under growing or highly variable stress, or where numerous resources within an ecosystem are utilized simultaneously, as in multispecies, multi-gear situations. Here this is illustrated for Malombe and Kariba (Boxes 5.1 and 5.2).
- One or a few large fluctuating resources – examples are clupeids and floodplain fisheries.
- Relatively small but numerous resources in stable environments – lakes, large reservoirs, near shore marine environments, or reefs.
- Systems of small interacting resources that fluctuate erratically – small lakes, ponds, lakes under highly variable abiotic stresses (anthropogenic or otherwise).

Assessment approaches related to the last three classes range from monitoring and stepwise forecasting, to the development of simple indicators addressing whole-ecosystem variables and models based on a community-level approach. In terms of research and regulations the cost of single-species management would become excessive for small fluctuating resources. Instead, management should concentrate on real-time decisions for the immediate future with expected outcomes based on continuous evaluation of time series of yields, catch rates or other indicators and possible explanatory variables (effort, water levels) of the fluctuating system in a comparative and stepwise management approach.

The complexity of the fisheries and the ecology and dynamics of the small interacting resources, exacerbated by the abiotically driven intra- and interannual variability, and the general small economic value of the stocks thus precludes use of models based on steady-state assumptions. However, even disregarding environmental variation, a number of generalizations can be made on the responses of multispecies fish assemblages to fishing that for all practical purposes may be used for predictions on reaction to stress. For instance, a systematic sequence of changes in a species composition as a response to increased stress has been observed on the North American Great Lakes (e.g. Rapport *et al.*, 1985 and Regier and Henderson, 1972) and termed the 'fishing down process'. Since this early work, much has been learnt from a diverse array of aquatic ecosystems. Welcomme (1999) presents a synthesis based on inland fisheries, and contributes to a framework for the development of indicators for management and conservation purposes based on the use of explanatory variables in a largely statistical context.

We will take these approaches as a starting point to address the differential impact of fishing mortality on fish communities that are subjected to different conditions of environmental stress. Based on the gradient between pulsed and constant systems emerging from our discussion of environmental drivers, we can use Regiers's dimension of temporal and spatial variability to typify the fish communities present in those systems in this dichotomy (based on Lowe-McConnel, 1987; Kolding, 1993 and expanded by Coulter, unpublished). This does not mean that the scale or extent of the systems is not important: we will return to that when discussing the dynamics of fishing effort later in this chapter. After discussing in some detail the character of fish assemblages in constant or variable systems, and the generalized reaction of fish communities to stress (see also the expanded theoretical considerations in the Appendix), we will discuss how a multispecies fishery may interact with these communities. We will then develop a conceptual model in which all of this information can be summarized in the context of the gradient between variable and constant systems, and which could be used to identify variables that can be used as indicators of change as a result of increased fishing mortality.

African freshwater systems differ in their temporal variability (Table 5.3) and can be arranged in two broad groupings: pulsed and constant systems. This arrangement of course represents extremes in a continuous gradient, and various systems will occupy different positions along this dimension. Furthermore, larger freshwater systems may have subsystems with either characteristic. For instance, Lake Tanganyika and other African Great Lakes have a pelagic ecosystem that is highly variable, while various other parts of the lake system have constant characteristics (Zwieten 2002c; Allison *et al.*, 1995). Variable systems encompass a large range of environments, many of which are associated with large African rivers: floodplains, swamps and allotrophic riverine lakes. Lake Mweru and Lake Kariba fall into the latter category. Furthermore, the class of variable systems includes endorheic lakes (i.e. lakes that have no river outlet) – e.g. Lake Chilwa and Turkana – or those that are parts of larger constant systems. Lake Malombe, for example, is grouped here as a constant system because it is a satellite of Lake Malawi's South-East arm, with which it has many characteristics in common. Littoral sections of the Great Lakes are considered constant, though the littoral is seasonally most affected by runoff from land. FAO (1993) asserts that the productivity of Lake Malombe is probably highly affected by runoff, while, as we have seen, in terms of water-level fluctuations, the system is much more constant than most of the other small systems discussed.

Table 5.3 *Examples of (seasonally) pulsed and constant African freshwater systems, with a comparison of typical fish community attributes and the implications for fishery in these systems. P/B ratio = Production/Biomass; F/Z ratio = Fishing mortality/Total mortality, where Total mortality (Z) = natural mortality (M) + fishing mortality (F). M2/Z ratio is Predation mortality/Total mortality. The arrow indicates that the system characteristics are a gradient over the dichotomy. ++ = high, rapid, many, long; --- = low, slow, few, short.*

Constant (a-seasonal) systems ⟶	**Pulsed (seasonal) systems**
Stable environments with mainly internal energy pathways	Unstable environments governed by pulses of nutrients
Littoral, benthic and bathy-pelagic zones of lakes Tanganyika and Malawi	River floodplains (e.g. Zambezi, Congo, Niger, Nile)
Lake Victoria	Swamps (e.g. Bangweulu, Okavango, Luapula, Cuvette Centrale)
Lake George?	Allothrophic riverine lakes (e.g. Turkana, Mweru, Malombe?, Kariba, Kyoga)
	Cyclic endorheic lakes (Chilwa, Chad, Rukwa)
	Pelagic mixing zones of deep lakes (Tanganyika)

Resource character		**Constant systems**	**Pulsed systems**
Biodiversity	Diversity	++	--
	Trophic groups	++ (specialized trophic pathways)	-- (short trophic pathways)
Dispersal	Migrations	--	++ (lateral/longitudinal)
	Mobility	-- (territorial; stenotopic)	++ (colonisers; eurytopic)
Life history	Life cycles	++	--
	Spawning	Continuous	Seasonal
	Fecundity	Reduced - parental care	High - no parental care
	Selectivity	K-selective	r-selective
	Growth to maturity	--	++ (mostly 1 – 2 years)
Population	Natural mortality rate	Stable	Fluctuating
	Predictability of mortality	++ (little variability)	-- (stochastic events)
	M2/Z ratio	++	--
	Biomass	++	--
	Productivity (P/B ratio)	--	++

Fishery implications	**Constant systems**	**Pulsed systems**
Exploitation rate (F/Z) can be	--	++
Regenerative capacity is	-- (fragile)	++ (resilient)
Yield potential is	--	++
Susceptibility to increased fishing mortality (F) is	++	--
Interannual variability in catch is	--	++

Table based on (Lowe-McConnell, 1987; Kolding, 1993 and Coulter, unpublished).

The typical environmental variability experienced by a system is reflected in the biological responses of fish species to it. The various attributes of fish species presented in Table 5.3 represent generalized life history characteristics found in a community typical for each class of environments. This resource characteristic is more or less determined by the 'ruling pattern of mortality' (Appendix section 2). In other words, the probability of dying from either abiotic or biotic causes determines the typical assemblage of fish species in a system. Under environmentally stable conditions, biotic causes such as competition and predation will prevail, and organisms will tend to develop mechanisms to escape mortality from other organisms (grow big, parental care, specialize), leading to the attributes listed under constant systems. If, however, the chances of dying are more unpredictable in terms of abiotic disturbances (floods, droughts, fire, etc.) then mortality will hit all size groups equally. In this case it could be a disadvantage to be a big long-lived fish and an advantage to be a small short-lived fish with attributes for a rapid recolonizing as listed under pulsed systems[23]. The basic question of whether fish yield is determined mainly by abiotic or biotic factors in these systems can now tentatively be answered, at least on a generalized system level. Fishery implications, that is the effect on the interaction between resource character and fishery, can be generalized over the two classes of systems[24] in five dichotomies (Table 5.3). Where the proportion of fishing mortality is low compared to the predation mortality from the community, and the biological turnover (P/B ratio, see Appendix section 3) is high, growing fishing effort is less likely to lead to overfishing or stock collapse. Species and communities having these characteristics have adapted to a high natural mortality, and possess a high regenerative capacity, so the susceptibility to increased fishing mortality is low. As fishing mortality can grow without much impact and the natural turnover is fast, the yield potential of these species is high. However, interannual variability in the landings of a fishery will also be high in these instances. Examples are: *Limnothrissa* (Kapenta) in Kariba, *Microthrissa* (Chisense) in Mweru and *Barbus* (Matemba) in Chilwa.

However, regardless of environmental variation, all fish communities contain species or segments of populations that have a differential vulnerability to exploitation. We distinguish three categories of susceptibility to fishing. Large groups of fish species and segments of fish communities can be fitted into these three groups based on rather diverse considerations of ecological and behavioural specialization and size characteristics (see Table 5.4). Under intensifying fishing pressure, which initially typically targets the large individuals in a community, the composition of a catch shifts (Regier, 1973; Hoggarth *et al.*, 1999a, 1999b; Welcomme, 1999). In general, in the successive development of a fishery, the more susceptible components of a fish community will give way to the more resilient species. Susceptibility here is considered as an intrinsic characteristic of a species or ecological group, by which we mean the relative ease with which they can be subjected to high exploitation rates.

The spawning concentrations of species migrating up-river are very susceptible to fishing pressure. Examples of rapid declines in stocks of such species abound. In our case studies, the example comes from Lake Mweru where the large cyprinid *Labeo altivelis*, the mpumbu, disappeared quickly from the fishery when demand for 'caviar' was high in the 1950s and spawning migrations were heavily hit (Kimpe, 1964). Spawning runs of the cyprinids of Lake Tana, Ethiopia are presently endangered through fishing (Nagelkerke *et al.*, 1995; M. de Graaf, pers.com.). Older age classes of all species generally disappear fast from a fishery in the early stages of its development.

---

[23] However, bigger specimens may be able to escape the vagaries of abiotic variation as well by enduring mechanisms such as hibernation or migration.
[24] But not over each and every species within these systems – see below.

Table 5.4 *Resource character and susceptibility to fishing of various components of a fish community.*

Susceptible ──────▶	Resilient ──────▶	Most-resilient
Old segments of populations of particular long-lived species (*Clarias* spp., *Lates* spp., *Hydrocynus* spp.)	Relatively unspecialized ecologically flexible species, widely distributed in rivers as well as in lakes adapted to fluctuating environments e.g. the fish faunas of Lakes Mweru, Chad, Turkana, Bangweulu, Chilwa	Small species with high population turnover rates (P/B ratios). Clupeidae such as *Limnonothrissa* (Tanganyika, Kariba, Cabora-Bassa), *Microthrissa* (Mweru) and cyprinidae such as *Rastrineobola* (Victoria), *Neobola* (Bangweulu), *Engraulycypris* (Malawi) and *Barbus* (Chilwa).
Species with longitudinal riverine migrations resulting in spawning concentrations (potamodromous species)	Tilapias, usually dominant in African fisheries, along with many species from the families of Alestidae, and catfishes (Clariidae, Bagridae, Siluridae and Mochokidae)	
Highly specialized endemics e.g. many (territorial) cichlid species of Lakes Malawi, Tanganyika and Victoria		

Based on Coulter, unpublished.

Large specimens of *Lates* sp. disappeared quickly from the catch both in Lake Turkana (Kolding, 1995) and in Lake Tanganyika (Coulter, 1991). Large tigerfish (*Hydrocynus*) and catfishes (Clariidae; Bagridae) also became rare in the catch in Mweru (Kimpe, 1964; Zwieten *et al.*, 2003b) and Kariba (Kolding *et al.*, 2003a).

The species of the intermediate category are more resilient to intensified fishing pressure. They have a wide distribution in rivers and lakes, and in many cases furnish the bulk of the yield of African freshwater fisheries together with the fisheries on small pelagic species. They are ecologically flexible, are relatively unspecialized in food preferences and behavioural patterns and are able to withstand high natural mortalities while living under highly variable abiotic conditions, such as in the swamps of Lakes Chilwa, Bangweulu and Mweru. The reproductive capacity of many of these species is high. Although many Tilapia-like species have rather specialized spawning behaviour and exhibit parental care (mouthbrooders, nest builders) they are very productive and can withstand highly varying ecological conditions[25] through adaptations such as stunting[26]. Narrow trophic specialization is generally not present: many species have a varied diet and are unspecialized omnivores, detrivores, herbivores or opportunistic piscivores (Bowen, 1988).

---

[25] Parental care is also a function of variable environment and circumstances of low oxygen content.
[26] The phenotypic ability to start reproducing at a much smaller size than would be the case under more favourable conditions.

The third category is species that are highly resilient to fishing pressure. These are the small pelagic species that have become important in the African fisheries in the past three or four decades. Table 5.4 lists these species, that all have high P/B ratios (see Appendix section 3) and high natural mortalities, which together with the 'down fishing' of their predators, enables them to withstand high exploitation pressures.

Total yields in multispecies fisheries are, in general, surprisingly stable over a wide range of fishing effort[27], though such stability may often be obscured by interannual variation in environmentally driven systems. This overall stability can be explained by the effect of aggregation of catches over numerous species, which reduces variability. In a recent discussion on the stabilizing effect of bio-diversity on the resource outcome, this phenomenon has been called the 'portfolio-effect', in reference to the stabilizing effect of a more diverse investment strategy on the stock market (Tilman et al., 1997; Doak et al., 1998; Tilman et al., 1998; Lehman and Tilman, 2000; Densen 2001). However, as different segments of a fish-community have a differential susceptibility to fishing and other stress, it has long been recognized that fish communities tend to react in broadly predictable ways to increased stress (Regier and Henderson, 1972; Regier, 1973 Rapport et al., 1985). The succession in exploitation of more susceptible to more resilient components of a fish assemblage under increased fishing pressure is a well known phenomenon, and has been documented for a range of freshwater and marine systems (e.g. Regier, 1973 and references in Welcomme, 1999; Zwieten et al., 2003b). An expectation that can be derived from these generalized observations is that the size of species caught in a multispecies fishery will reduce over time, while the species composition of the catch will change to components of an assemblage that can withstand higher mortalities. This process has also been linked to the trophic level of each of the components of a fishery, and an annual mean trophic level of the total catch of a fishery can be calculated. A downward trend in the mean trophic level indicates that a fishery is moving down the food web as new small-sized resources are increasingly being exploited, while the large-sized species are reduced in abundance.

Not only biological characteristics intrinsic to a species determine susceptibility to fishing pressure. Other factors are changing environmental conditions and the susceptibility of species to specific gears: in other words the accessibility of stocks to the gears in use. Changing environmental conditions may change susceptibility to fishing, by changing the accessibility to gears. Drying pools in floodplains where fish are concentrated and easily fished out are an obvious illustration of this effect. Conversely, high water levels in floodplains and swamps will provide more space and habitat for shelter, and catch rates will decline because of the resulting reduced accessibility of fish to gears. In Lake Chilwa, under lowered water conditions, all species become more susceptible to fishing without a change in fishing methods (Zwieten and Njaya, 2003). Similarly, in Lake Mweru water draw down after the seasonal peak in May results in extremely high catch rates of *Oreochromis mweruensis* in the shallow lake area immediately adjacent to the swamps, while catches of this species are reduced under seasonal flood conditions.

---

[27] This has also been found in larger ecosystems such as the North Sea with all fisheries combined.

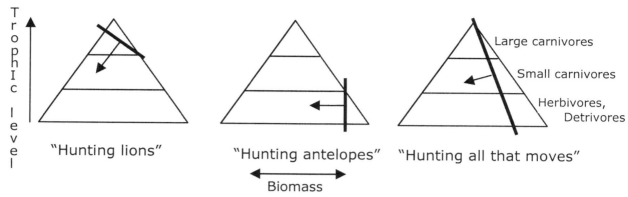

Figure 5.5 *The trophic levels in a community at which a fishery intervenes. Triangles represent trophic pyramids of animal communities with predators at the apex and animals feeding on primary production and detritus at the bottom. The width of the triangle at any level represents the relative biomass of that level. Black lines represent selective exploitation, arrows the direction of increased pressure. The three triangles could each represent a different fishery, for example: a sport fishery on tigerfish ("Hunting lions"), a gillnet fishery on tilapiine fishes such as the Oreochromis fishery in Mweru comparable to grazers in wildlife ("Hunting antelopes") and a fully developed fishery in which all trophic levels are harvested proportionally to their biomass ("Hunting all that moves").*

The accessibility of a stock to a gear is related to catchability and selectivity (see Chapter 2). Considering that all fishing gears are selective means that each gear affects only a certain portion of a fish community, that is the fishable stock. Basically there are two types of fisheries: those that employ only one gear, usually gillnets, and those in which a high diversity of gears are employed. As a consequence, the intervention of a fishery will either be selective to certain components of a fish assemblage or will to a greater or lesser extent target all components present. Sport fisheries on predatory fish such as *Hydrocynus vittatus* in Lake Kariba, commercial fisheries on *Lates niloticus* in Lake Victoria or on large *Lates* spp. in the early stage of the pelagic purse seine fishery of Lake Tanganyika (see Box 5.3) are examples of the first type of fishery. A second example of this type, but with a completely different impact, is the early stages of gillnet fisheries of Lake Mweru or Malombe, when larger mesh sizes were used that mainly (though not exclusively) targeted large *Oreochromis* species. The difference between the two examples is that the fishery intervenes on entirely different trophic levels within an ecosystem – predators vs. detrivores/herbivores (Figure 5.5). Lake Mweru also gives the example of a developing fishery diversifying into targeting successively more components of the fish community. Mesh size in stationary gillnets decreases, while more and more active methods are employed, such as seining (open water and beach). Lastly, fish attraction through lights on the pelagic species *Microthrissa* (zooplanktivore) or through Fish Aggregating Devices (FAD's) on *Alestes macrophthalmus* (facultative piscivore/insectivore) develops.

The result is a maturing small-scale and subsistence fishery, in which many components of the assemblage are targeted, without any of them disappearing from the fishery, though some species reduce to low stock levels. Thus, the conceived negative image of "fishing down the food webs" is not necessarily ecologically bad, but just represents a fishing pattern illustrated in Figure 5.5 where the fishery develops to exploit all trophic levels – hunting everything in proportion to the natural P/B ratios. In principle, such a fishing pattern could be much more harmonious for the overall natural mortality pattern than a selective fishery with gear and mesh size regulations (Kolding, 1994; Misund *et al.*, in press). Floodplain fisheries in the large Asian rivers, for example the Mekong (Coates, 2001) are examples of highly complex fisheries in which a large number of species are targeted by an enormous diversity of gears, without evidence for any of these species disappearing from the system. The Bangweulu case study (Kolding et al., 2003b) is a similar example of how the different gear types (many of which are illegal) are targeting different parts of the fish community.

To understand the complex simultaneous effects of changing environment, fisheries impacts and trophic interactions (predation and competition) on the biomass and structure of a fish community, we will use a conceptual model that is derived from an approach used to analyse properties of whole ecosystems based on the distribution of biomass over the size of organisms (Sheldon et al., 1972). By including information on externally driven productivity on the one hand, and size-selective stress on the other hand, changes in the shape of the biomass-size distribution may indicate on a 'whole' system level what size and extent of effects these have on this particular group or assemblage.

As there is generally a larger biomass of small fish (small species and juveniles of large species) compared to large fish, the overall shape of a biomass-size distribution is a descending curve over size (Figure 5.6). Variations in the shape of this curve, the slope and the intercept with the vertical (biomass) axis indicate systematic changes in the size structure of communities that can be related to the mortality pattern (see Appendix 1) and variables such as mean lake depth, lake productivity, water turnover rate and human impacts (Cyr et al., 1997). Variations in the intercept with the vertical axis indicate that average overall biomass in the system is a function of the availability of nutrients to the system, rather than a function of the structure of the particular biological community (Boudreau and Dickie, 1992).

Changes in the slope indicate changes in the size structure of the group or assemblage. The approaches derived from these insights deserve special attention in tropical fisheries, as they could provide promising ways to address the problems related to the assessment of multispecies resources. The size structure of a multispecies catch can provide a global interpretation of a fishery (Gobert, 1994) and relations between size structure and impact parameters can be used at a local scale to describe the size structure of freshwater communities. Though much work still needs to be done to turn this tool into a method of formal assessment of a fish community, it can already be used as a way of visualizing a large amount of complex information (Zwieten et al., 2003b). In this way, the information contained in the length structure of catches, nowadays extensively sampled in many small-scale fisheries, can be used to their full potential.

Size information tells us something about the species susceptibility to fishing. Due to the selectivity of most fishing gears (usually regulated in terms of the prohibition of fishing on small individuals), it can in general be said that highly susceptible large fishes are on the right – and those not so susceptible are on the left in a biomass-size distribution. Figure 5.6 visualizes the impact of the magnitude of the seasonal pulse on various parts of a fish community, which has been described as "riding the wave of annual productivity as it rolls through the extended size spectrum from phytoplankton to big fish" (Pope et al., 1994). The impact of this 'wave of productivity' is dampened along the size spectrum by individual life history traits of species and size groups (growth and mortality rates, that is productivity, generally decreases with increased size) and species interactions at higher levels such as predation and competition. In constant systems, the seasonal energy pulse, which acts on the smaller sizes of the fish community first, is much smaller. Together with the long-term stability of these environments, this allows for the development of life history strategies that result in narrow niche specificity (stenotopy, territoriality, trophic specialization). These adaptations towards efficiency rather than proliferation make susceptibility of fishing for these smaller species also higher. Variability in standing biomass of smaller-sized individuals is higher than larger-sized individuals, but again larger in pulsed systems compared to constant systems. Energy distribution, regulating the abundance through interactions of competition and predation, is larger. Lastly, Figure 5.6 indicates that the range in size of fish in constant systems is generally larger. In summary, the alternative life history traits of species as a function of different selective forces in the pulsed or constant environments, results in different biomass size-distributions. The biological attributes and implications for exploitation are the same as listed in Table 5.3.

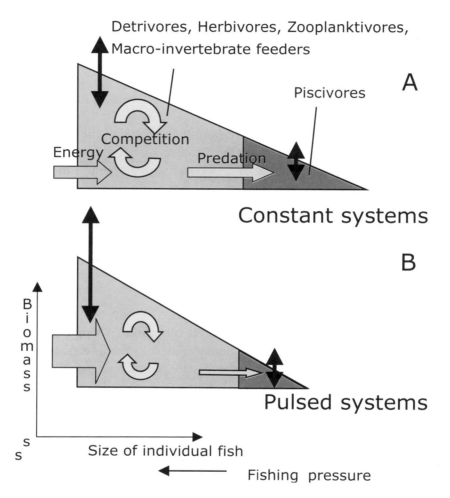

Figure 5.6 *Biomass-size distributions, variability, and energy pathways in A) constant and B) pulsed systems. Triangles characterize fish communities by biomass and size while variability (vertical arrows) and energy flow (horizontal arrows) indicate dominant mortality patterns. Biomass decreases with fish size and predatory fish (blue triangles) are generally larger than their prey. Energy flow through pathways starts from seasonal input of nutrients resulting in seasonal changes in productivity (red arrows). Energy in a fish community is partitioned through competition and predation (green and blue arrows). Variation in biomass caused by changes in energy input is larger with smaller-sized fish (black arrows). Increasing fishing pressure generally results in decrease in biomass of large fish and increased catches of smaller fish.*

The impact of increased fishing effort is, in general, described as decreasing from large to small specimen of fish. However, as indicated in the discussion on the selectivity of fisheries and the preference of certain target species, it will be clear that the situation may be more complex. For instance, in Lake Mweru the decreased mesh size of the dominant fishery on the lake, the gillnet fishery, did reduce the size of fish in the catch but also targeted different sections of the fish community, and thus released pressure on the population of larger *Oreochromis* (Zwieten et al., 2003b). In the following section we will discuss the dynamics in fishing effort in a context of scale of the fishery, the level of externally driven variability of the system and its consequences for the fish communities.

5.5 Selectivity and scale of operation of fishing patterns: dynamics of fishing effort

Previous chapters emphasize and explain the small-scale character of fisheries in SADC freshwaters. "Small-scale" fisheries generally operate within geographically limited areas and intrinsically depend on local resources. This is in contrast to technologically and organizationally more developed fisheries with a broader spectrum of options in terms of fishing grounds, markets and alternative investment opportunities (Thomson, 1980; Panayotou, 1982; Bâcle and Cecil, 1989; Misund et al., in press). Since 1950, African freshwater fisheries have grown much more rapidly in the small-scale sector than in the technologically more advanced sectors, which have not thrived very well (see Chapter 3). How problematic is this growth in technologically less advanced but numerically abundant effort from a biological point of view, in particular in pulsed systems? In all our case studies, increased effort, with annual rates of increase of numbers of fishermen between 2% (Mweru) to 6% (Chilwa) had a significant effect on stocks (Table 5.1). However, only in one case, Malombe, has increased effort led to a collapse of an important stock. Increased effort in Mweru and Chilwa was mainly population-driven, but in Malombe growth of effort was concluded to be investment-driven. Does this mean that a population-driven increased effort, operated with a high variety of gears on a small-scale is biologically less harmful than increased effort as a result of accelerated investments in more effective technology in these pulsed systems?

Firstly, observed effort in terms of density of fishers is not uniformly distributed across the different systems. The average allocation of effort as number of fishermen per unit area in the various African freshwater systems is roughly correlated with the magnitude of the seasonal pulse into the systems as quantified by the RLLF-s (Figure 5.6A). In turn this could be a function of the overall high productivity of these systems (Figure 5.6B). Though catch rates and effort are not independent, this could indicate that the more pulsed systems in general support a higher effort than the more constant systems. There is little evidence from our case studies that this high effort has had a negative impact on the regenerative capacity of most of the stocks. This result supports the theoretical fishery implications for pulsed systems outlined in Table 5.2.

Secondly, to understand the dynamics of fishing effort within each system – i.e. the resultant effect of the choices made by all fishermen – it is important to look at the scale on which a gear is operated relative to the variability in abundance of the resource it targets. It is the scale of operation that makes a small-scale fishery distinct from technologically more advanced fisheries and determines the investment level and amount of labour needed (Misund et al., in press). The scale at which a fisherman can operate his gear determines how the variability of the resource will be reflected in his day-to-day catch. One could also say, it is the way a fisherman uses his gear to aggregate fish that determines how his daily catches are stabilized, as well as the effect it has on the stocks.

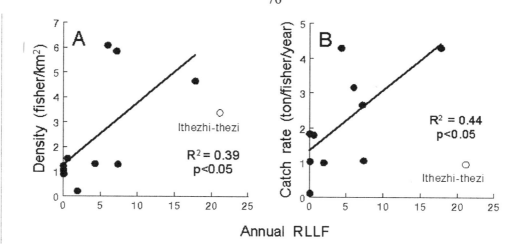

Figure 5.7 *The relationship between annual relative lake level fluctuations (RLLF) and effort density (A) or catch rates (B) in 11 African freshwater systems. Lake Itezhi-tezhi is not included in the regression lines. Data for the number of fishers and landings are given in Table 2.2, data for RLLF are given in Table 5.1 and Kivu = 0.06, Turkana = 1.97.*

If a fisherman wants to stabilize his income from fishing through stabilizing his daily catch he has three options (or combinations of options):

- *Option 1*: invest in more, larger or more efficient gear, in other words aggregate fish in space,
- *Option 2*: use the gear more intensively (extra labour), or aggregate fish in time,
- *Option 3*: diversify to other species and be less dependent on one resource, in other words aggregate fish in the fish community through catching a diverse assortment of fish.

Lastly, he also can stabilize income by doing something else besides fishing (Aarnink, 1997; Oostenbrugge *et al.*, in prep.; Zwieten *et al.*, 2003b).

Thus a high diversity of species in a catch will also reduce overall variability, even if the catch of each individual species will be low or highly variable. Most important regarding the effects of the choice of these different options in effort allocation, is that the variability in fish stocks, both in total biomass and in relative abundance of the component species, is determined, as argued before, by longer-term (interannual, decadal) changes in annual flood pulses. In other words, both the fishing pattern and the resource variability will result in a more or less variable outcome. Both are important variables for a characterization of the dynamics of effort in a fishery (Appendix section 5).

Small-scale fishing gear can be classified in three development categories with increasing scale of operation and more or less with increasing level of investment (Misund *et al.*, in press):

Traditional fishing implements are locally produced and are operated from small canoes or rafts and by wading or diving fishers. Simple hook and line and handline, spears, harpoons, tongs, pots, traps, dip-nets, gillnets, small beach seines, weirs and barriers are included in this group. Intermediate fishing implements are nets (gillnets, seines and lift-nets) and much of the hook-and-line gear that are factory produced and made of synthetic materials.

Modern fishing implements are trawls and other mechanized active gears, including the mechanized haulers that increase the spatial coverage of a fishery. Modern accessories to fishing such as hydro-acoustic equipment (echo-sounders), and global positioning systems increase the efficiency of the fishing operation by increasing the spatial window over which the fish can be searched and found or that will give accurate site information.

Most African fisheries contain a mix of the first two groups of fishing implements, the first being associated more with riverine and floodplain fisheries, while the second are dominant in lakes. Modern fishing implements are rarely found in African freshwater fisheries. Increase in efficiency is generally reached by an increase in the scale of operation through more intensive use of gear – for instance through active methods such as seining or 'beat' fishing – and not by auxiliary technology. These two development categories with their lower capital investments and reduced maintenance costs that are used with simpler fishing methods, make small-scale fishers more adaptable to the exploitation of a multispecies stock that show large changes in stock levels of different species. In other words, in African freshwaters small-scale fishermen often opt to stabilize catches by increasing the number of species caught or switching between them by using a high variety of capture methods.

This adaptability could be seen as utilizing multispecies stock in a more efficient and biologically more sound way, compared to technologically more advanced or larger scale fisheries that are less adaptable to changing circumstances. Some of the following characteristics typical of the fishing patterns of small-scale freshwater fisheries, are indicative of adaptability:
Effort (overall fishing pressure) is unevenly distributed among different systems. Pulsed systems in general support a higher fishing intensity than constant systems suggesting that the productivity of the system to a certain degree is regulating the effort despite the overall population-driven growth (Figure 5.7)

Allocation of effort is unevenly distributed in time and space within a system. For example in Mweru, most fishers are concentrated in the swamps, lagoons and the highly productive southern part of the lake and effort is to a large extent a function of the physical environment of the coast, access to land and to markets. Fishing is hardly ever the sole source of income (see Chapter 3). Effort allocation is seasonal during periods of receding water levels and associated longitudinal or lateral migrations of fish, while weather conditions highly affect spatial and temporal fishing patterns on the pelagic fishery (Zwieten *et al.*, 2003b). In Kariba, most fishing effort is in the inshore, in Zimbabwe on the eastern part of the lake around the main population centres. Lastly, migrational shifts in effort take place both within a fishery and in and out of a fishery (see Chapter 3 and Zwieten and Njaya, 2003; Zwieten *et al.*, 2003b; Overå, 2003).

Fishing patterns are highly diverse and *often labour-intensive*: numerous people are active with an extraordinary variety of specialized capture solutions to different resources, environments and seasons resulting in labour-intensive multi-gear/multispecies exploitation patterns. This is not typical for African freshwater fisheries, but has been described for many situations in tropical freshwater and coastal systems (von Brandt, 1959, 1984; Nédélec, 1975; Nédélec and Prado, 1990; Coates, 2002; Christensen, 1993; Hoggarth and Utomo, 1994; Hoggarth *et al.*, 1999a, b; Zwieten, 2002d; Medley *et al.*, 1993; Misund *et al.*, in press). River and floodplain fisheries often have limited numbers and sizes of intermediate fishing implements per fisherman. For example, in Mweru swamp fishermen have on average one gillnet compared those in the lake that have between 10-20 gillnets of the same size each (Zwieten *et al.*, 1995). Gillnets and other intermediate fishing methods are used in most lakes and reservoirs for inshore and demersal species, often as both active and passive fishing techniques. The offshore and pelagic regions of these fisheries have become more important over the past three decades, where small pelagic species are caught in more specialized nets often in combination with light attraction.

These complicated fishing patterns and highly diverse fishing methods operated on a small scale, in general have a small daily output – a bucketful – for each individual fisherman, and each capture device is intrinsically associated with specific selectivity. In other words, the impact of fishing is distributed over large numbers of fishing units and all methods have a differential impact on the ecosystem. Much attention and many legal regulations are focused on the supposedly detrimental effects of the so-called less selective gears such as seines, small mesh sizes in gillnets, drive- or beat fishing, barriers and weirs. However, in practice little is known about the actual impact of these methods. In the few instances where the actual impact of these gears has been studied, it is an open question how "detrimental" they actually are (Kolding *et al.*, 1996; Chanda, 1998; Bennett, 1993; Clark *et al.*, 1994 a, b; Lamberth *et al.*, 1995 a, b, c). Small-meshed beach-seines on small barbs and juvenile bream have been used for a long time in Chilwa without much apparent negative effect (Zwieten and Njaya, 2003), though by contrast the large beach-seines used in Malombe were highly detrimental to the stocks of *Oreochromis*. Furthermore, as many gears can be utilized in variety of ways and while the lifespan of most fishing gears is generally low, fisheries can readily react to changing circumstances and seasons. For example, in Lake Mweru gillnets are used in at least eight different fishing patterns: as stationary gears, in various types of open water and shoreline seining, drift-netting and in combination with other gears such as knobbed sticks for 'beat fishing' and Fish Aggregating Devices, all targeting different subsets of species (Zwieten *et al.* 2002b). Traps and gillnets normally do not last longer than two to four years and shifts in mesh sizes rapidly take place over a whole fishery.

As a result of these characteristic effort dynamics, multi-gear/multispecies fisheries, in particular in highly variable ecosystems such as floodplains, could be producing an overall species abundance and size composition that closely matches the size and species structure of a fish community. In principle, a fishery that harvests all species at all trophic levels at rates proportional to their natural mortality pattern during their lifespan will be non-selective on an ecosystem level. Non-selective overall harvesting patterns would conserve the fish community (see Appendix section 4) as the relative structure of the ecosystem would be maintained, only each component would be smaller (see Figure 5.5 "Hunting all that moves"). For instance, the Mekong floodplain fisheries seem to have persisted unchanged (although with high temporal fluctuations) with very high numbers of people fishing with a high variety of methods for as long as our observations can tell. Even the susceptible migratory "white fish" and predatory species still form important parts of the catch, while giant catfish still are present. Where biodiversity has been adversely affected, it is generally where ecosystem integrity was not maintained (Coates, 2002). In other Asian river systems, such as the Bramaputhra and the Ganges, migratory fish have disappeared and fishing is now completely dependent on small individuals, indicating that eventually adverse effects may occur (Hoggarth *et al.*, 1999a,b).

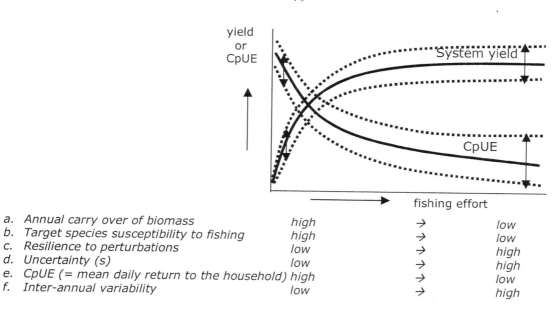

Figure 5.8 *Generalized development of fishing yield and catch rate of a fishery with increasing effort. The broken lines represent increasing variation around the mean yield and catch rates over time (vertical arrows). The six characteristics in the table below the plot change with increased effort as indicated (high/low). They refer to total fish biomass of a system (a, b, c and f) and to the outcome of the fishery (d, e and f).*

However, this does not mean that increasing effort is without problems. Increased effort, population-driven or otherwise, invariably leads to decreasing catches per fishing unit (CpUE) (Figure 5.8). Some fish species, found mainly but not exclusively in constant systems, will be more susceptible to fishing pressure (Tables 5.3, 5.4). As an overall harvesting pattern of a fish community where the fishing mortality is strictly proportional to the natural mortality of its constituent species probably does not exist, selection will take place. Increased reliance on the smaller species or the smaller, juvenile, segments of a fish community leads to increased variability in the catch of individual species[28]. This is caused both by the natural mortality patterns of the smaller species as they follow the boom and bust periods in variable systems, and as a result of a decreasing overall biomass. The time series of catch rates of the industrial fishery of Lake Tanganyika, described in Box 5.3 shows this, while this is also an example of option (1) of the choices for catch stabilization through spatial aggregation.

In Malombe, fishing with large beach seines on *Oreochromis* spp. resulted in average monthly catch rates that, though highly variable[29], showed no trend over the years. No effect of decreasing water levels was observed. In the same period, catch rates for *Oreochromis* in gill-nets exhibited fast declining trends and were significantly affected by changes in annual lake-levels, with expected time lags as fish "grow" into the gear. The fact that this effect was not seen at all in the large beach seines can only be explained if a change in the scale of operation – through a change in gear size, intensity of gear use or spatial allocation (options 1 and 2) – took place (Zwieten *et al.*, 2003a). In Malombe, stabilization of catches with these large seines has been obtained through increased spatial allocation of highly efficient labour-intensive gear. In other words, in this case increased scale of operation overrode the natural variability induced by flood pulses. As flood pulses have a much more limited effect in this system compared to the

---

[28] Though less so in the total daily multispecies catch (Oostenbrugge *et al.*, in prep.)
[29] In the data examined this was also due to an uncertain definition of effort.

other systems researched, stocks of both *Oreochromis* and the small haplochromine complex had limited opportunity to bounce back within the period of high fishing pressure.

Box 5.3 *Development in catch rates of the industrial pelagic fishery of Lake Tanganyika: one reason for its collapse*

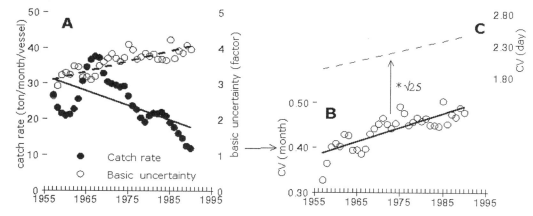

Figure 5.9 *A. Catch rates (closed circles and continuous line) and basic uncertainty (open circles and broken line) from 1956 to 1992 (five-year moving averages). Basic uncertainty is the variation excluding effects of the long-term trend and seasonality, expressed as a factor around the mean: for example the monthly average catch in 1990 was 11.7, the factor 4.0. So, catch rates varied 'randomly' over this five-year period between 2.9 and 46.8 ton/month/vessel (see Zwieten and Njaya, 2003); B. Monthly variability (CV) per vessel in catch rates of this fishery based on Figure 5.9A.; C. Estimate of the day-to-day variability of a single vessel, by multiplying the monthly CV by $\sqrt{n}$, where $n = 25$, is the average number of days per month a fishing vessel goes out.*

In Burundi on Lake Tanganyika a semi-industrial purse seine fishery using light fishing techniques utilized the highly productive pelagic fish community of the lake. Initially the fishery targeted larger-sized predatory fish, but after 1956 it switched to small fish, two clupeid species, and their smaller predators. Over time, a large artisanal fishery developed, targeting the same resource (Zwieten *et al.*, 2002c). The demise of the semi-industrial fishery was caused not only by declining profitability but also by their reliance on a single resource with a highly uncertain outcome and no means to increase their technological capacity to overcome these uncertainties. Over the years, the scale of operation of an individual industrial fisherman remained the same, with limited variation in total effort, while the total mortality on the pelagic stocks increased due to the increase in artisanal fishing operations. This changed the encounter rate of the individual operation. Variability, already high 1957, increased after the switch from the susceptible large top predators to its prey, the freshwater herrings, around 1958 and kept on increasing to extremely high values two decades later[30]. Limited in space, being on a lake, while highly dependent and switching between only a few species, the only option for the industrial fishermen to reduce variability and stabilize catches was to invest in more technology. This did not happen and the fishery stopped. It can be expected that an artisanal fisherman, operating with smaller units has an even higher variability. For fisheries to sustain such high uncertainty (in daily income!) means that back-up systems are needed to stabilize income or maintain at least a partially low but certain income to avoid the risk of poverty (Oostenbrugge *et al.*, in prep.).

---

[30] This daily variability is extremely high: by comparison, a small-scale gillnet fisherman generally experiences a daily detrended and deseasonalized CV = 50% – 70% (Densen, 2001; Zwieten, *et al.* 2002a).

The larger *Oreochromis* had little chance to occasionally produce large year classes. In Malombe thus a mismatch exists between the natural variability in fish production and the scale of exploitation.

Lastly, an example of Option 3: in Lake Mweru the effect of decreased catch rates and increased variability has resulted in an increasing diversification of fishing methods and target species, as previously unexploited stocks are now targeted. The shift to smaller mesh sizes has resulted in a higher diversity in the catch. In this case, increasing the number of species in the catch reduces variability in individual catches. Each of these species may have highly different stock sizes, since the opportunities to reproduce and the abiotically induced mortality patterns differ both within and between years and are different between species. Timing, duration and extent of flooding are important factors in this dynamic. For instance, large variation in water levels has resulted in strong year-classes of *Oreochromis*, which could partially escape the gillnet fishery in inaccessible areas during flooding. The real danger for these species would be if the occasional larger flood pulses did not take place over a long period of time. As with lake Chilwa, where flood pulses are reflected in the catch in the same year or one year later, increased effort in Mweru will mean that boom-bust periods will be seen in the catch of individual fishermen with much shorter time lags than previously when larger fish were targeted.

From this analysis we can conclude that:

Artisanal small-scale fisheries such as in Lake Mweru, by hedging the inherent variability in relative abundance of multispecies stocks, and opting to target many species simultaneously, are developing an overall fishing pattern that could in principle conserve the ecosystem. On the other hand, where operations override the inherent variability by increasing scale and maintaining catch rates at the same level, this will lead to problems.

In other words, the multi-gear (overall unselective) fishing pattern employed in many small-scale fisheries existing in pulsed systems, combined with the ability of fishermen to rapidly change their target, could be seen as an ecologically optimal fishing pattern over a very large range of fishing pressures. In these cases, the conflicts arising from the 'management' implementation of existing gear restrictive regulations (based on single-species models and considerations), are in many situations, at least from an ecosystem perspective, largely futile.

In situations where multi-gear operations exist with limited investment per operation many people in principle will be able to start fishing. At several points in our analysis so far we have seen indications that effort levels are dependent both on productivity and changes in productivity. Increased effort will invariably mean that catch rates will go down, while at the same time the inherent variability of pulsed systems can produce boom-and-bust situations. In that case, catch rates could be a regulating factor for fishing effort by itself if other options to produce an income still exist. This is in contrast to the idea of Malthusian overfishing that predicts that declining catch rates pose a trap from which fishermen cannot escape.

5.6 Information on catches, effort and environment in African freshwaters

As both the Malombe and Kariba cases indicate, generating the classical models and obtaining basic population parameters such as growth and mortality have been useful in interpreting biological developments and status of particular fisheries. However, taken as a scientific point of reference on which management measures could be taken, which for a long time was considered the ideal in many African countries, the model outcomes only gave a misguided sense of controllability. The inadequacy of the models to encompass the effect of the fishing pattern on the whole community may also create disillusion if enforcement of measures taken is unsuccessful. For instance, in 1999 all fishing methods used daily in Lake Malombe were in fact forbidden (pers. com. Dr. Mapira, Director of Fisheries in Malawi)! Uncertainty about the effectiveness of the measures taken and about causal relationships only makes matters more complex.

To overcome the problem of information for fishery management in the face of these complexities, a pragmatic approach should be adopted in attempting to translate uncertainties into the decision-making process and management practice. The classical models do help in thinking about complex data sets, and in generating intuitions about a system. However, the next step, which is often forgotten, is how they could help in defining information needs. The results of models, however crude, are just one piece of the evidence in the translation of uncertain information into knowledge relevant to management (Zwieten *et al.*, 2002d). Observations outside the model invariably prove to be highly relevant to the issues at hand. Thus, learning about general system behaviour through evaluation of long-term information will aid in generating the necessary intuitions on what may or may not work. The important questions to be answered are: what is the gross abiotic and biotic behaviour of a system considered, what information can be obtained directly from evaluating time series of catch and effort data and other long-term data, to assess and predict changes in fish stocks in multispecies assemblages in response to fishing or other stress?

We have examined the evidence of long-term biological effects of increases in fishing effort, population- or investment-driven. Observed changes in fish stocks in the freshwater systems in question could be related to increased fishing pressure in all cases. But, size, extent and duration of flood pulses that act as environmental drivers are also invariably extremely important and cause many of these pulsed systems to be more or less resilient to increased effort. Growth in effort is discussed in the context of the scale of fishing operation relative to the variability of the ecosystem harvested, and the choices a fisherman could make to stabilize his catch. It has been shown that numerous small-scale operations using a variety of gears could conserve the fish community, and are probably less detrimental than increased effort as a result of increasing the individual scale of operation. For highly pulsed systems, the natural boom-and-bust will to a large extent regulate effort by itself, or as Beverton (1990) puts it "[some] fisheries cannot be driven to extinction because the fishermen will disappear before the fish". And even if they are intermittently 'overexploited' they have the capacity to regenerate fast when conditions are favourable. For the more stable systems, a higher potential danger of over-exploitation exists since they need more time to recover. For these systems, it may be appropriate to monitor the catch rates to learn their limits empirically and possibly intervene with effort-regulating measures. Most important, however, in the transition from traditional single-species classical models to a more comprehensive community approach taking more than 'effort' into consideration, is the need to learn more about individual systems' response to exploitation. From this perspective, the continued collection of catch, effort and environmental parameters is paramount.

The case studies on the five lakes aimed to address the possibility of detecting changes in fish stocks under increased effort and changing environmental conditions using the existing time series obtained through Catch and Effort Data Recording Systems (CEDRS), or experimental gillnet survey data, combined with time series on lake levels. It appears that much of the information present in time series of catch and effort monitoring and experimental surveys is severely under-utilized. Data obtained through the CEDRS's are in many situations often used solely to calculate total yields in statistical reports. Combined with effort data, attempts are sometimes made at calculating sustainable catch levels through formal stock-assessments, but with limited success. The information used in such models is often considered too unreliable. However, we have shown that information that could be derived from the time series of catch and effort is not limited to these applications. An analysis of trends and variability in catches, catch rates, fishing effort and water levels can produce empirical relationships, which in the proposed pragmatic approach can be used to predict not necessarily when something will happen, which is the aim of stock-assessments, for example, but what could happen if something changes (Kolding, 1994). This can only be achieved through continuous and mandatory evaluation of time-series information, also in relation to environmental changes, within the administrative structures responsible for management. Such evaluation will lead to knowledge of what can be perceived, while on the basis of this, expectations on the effectiveness of measures can be formulated. With an explicit analysis of uncertainties – also as a function of data-collection procedures – such trends can be qualified (Zwieten and Njaya, 2003; Zwieten et al., 2003a), and tell how fast changes occur and how fast the effect of management measures directed to stock protection will become visible (Densen, 2001). Examination of long-term experimental surveys can be used to comparatively examine changes in the fish communities, possibly in relation to changes in fishery or the environment (Karenge and Kolding, 1995a, b). From such results, important indicators can be derived which will aid in management decisions. In the formulation of information needs, there is thus a need for the 'long view' – the longer the time series of catch, catch rates, effort and general system indices such as water levels, the more can be learned about the behaviour of a fishery and its effect on the regeneration of stocks (Bjorstad and Grenfell, 2001; Yoccoz et al., 2001).

6. IMPLICATIONS FOR FISHERIES MANAGEMENT

6.1 Summary of findings

As we have shown in this report, the socio-political and ecological dynamics of the societies and lakes studied are quite complex. This complexity is not fully captured either by the assumptions underlying model-based management approaches exemplified by Common Property Theory, or by the neo-institutional framework underlying much of the co-management approaches. It is also becoming increasingly evident that the management measures implemented in the southern African freshwater fisheries are not adequately based on empirical knowledge.

In this chapter we will first summarize our main findings. Then we will discuss the implications of these findings for the development of a management approach in small- and medium-sized freshwater fisheries that takes both the natural and social complexity of the systems into account, at the same time as attempting to ensure equitable and sustainable utilization of the resources.

Even though great differences can be observed between ecosystems and countries, a number of lessons can be learned from our studies that have important implications for the way management and/or co-management is conceived and implemented in many African freshwater fisheries. The following are the main empirical findings from our studies of small- and medium-sized lakes in Zambia, Zimbabwe and Malawi:

- Growth in fishing effort in this region has mainly been population-driven.
- Environmental drivers, especially changes in water level, are most significant for variation in the productivity of fish stocks, and population-driven growth in fishing effort has not been too harmful for the productivity of the fish stocks in these lakes.

- There is considerable mobility of people in and out of the fishery sector.
- Population-driven growth in effort tends to increase when macro-economic conditions deteriorate and/or the biological productivity in a lake improves. Reductions in population-driven growth in effort take place when local access-regulating mechanisms exclude newcomers, and when investment-driven growth in effort results in rising entry costs for those who wish to join the fishery.
- Investment-driven growth rarely takes place in these fisheries: only two cases have been observed. When such processes occur, the investors are entrepreneurs from outside the fishing communities, or the investments are made possible by input of capital generated in sectors other than fisheries.
- Where investment-driven growth takes place it has led to the collapse of specific fish stocks.
- The main constraints on investment-driven growth of fishing effort in the southern African countries are that the local institutional landscape impedes the development of infrastructure and access to capital and markets.
- People's flexible adaptation to the ecological and economic environment through frequent entries into and exits out of the fishery sector, facilitates the function of SADC freshwater fisheries as a "safety valve" or, in other words, as a buffer against poverty.

6.2 What are the implications for management from a biological perspective?

How to manage fisheries effectively has been intensively debated for several decades. For quite some time the "classical approach" to fisheries management has leaned heavily on the prediction and estimation of optimal yield levels such as maximum sustainable yield (MSY), to which effort levels should be adjusted. Although models pertaining to this approach were initially developed for the management of large-scale fisheries on single stocks in relatively stable environments in temperate climates, it has become the dominating approach worldwide. However, setting optimal catch and effort levels on a species-by-species basis is highly problematic anywhere in tropical multispecies and multi-gear situations, and the information needs for such an approach in the small-scale SADC freshwater fisheries discussed here are prohibitively costly. Overall catches and generalized levels of effort (numbers of fishermen and gears) can be known, but the highly diverse and dynamic fishing patterns make the establishment of causal relations between catches and a specific level of effort virtually impossible. In many ways the approaches must be considered harmful to the development of management systems. Management proposals based on single-species model predictions give a strong incentive towards adopting rather drastic measures on effort control and gear regulations in situations where fishing may represent an important means of survival for many people. Moreover, the MSY approach has severe limitations in systems with large environmental variability, and for such systems it is extremely difficult to causally relate trends in fish stocks levels to fishing alone. Such trends are often unclear and hidden because of environmental variation acting over different timescales causing what is called in statistical terms "coloured noise" (Densen, 2001).

In fact, we have shown how environmental factors are often more significant than fishing effort in the explanation of changes in fish production in the lakes studied. In Chapter five it is argued that water levels are a key (environmental) driver explaining the productivity of fish stocks. Productivity in many of the smaller freshwater lakes is extremely high, and for various reasons many stocks are highly resilient to increased pressure and have the capacity to bounce back if the pressure is released. Lake Chilwa is an extreme example as it has dried up completely six times within the last 150 years, but as soon as the lake has filled up again, fish stocks have recovered. Another important observation in numerous multispecies fisheries, including some of the systems discussed here, is that levels of total yields appear to be surprisingly stable over a large range of fishing effort. Underneath this relative stability of total yields, changes in species and size composition occur, both as a result of fishing and as a result of environmentally driven processes. Lastly, fluctuations in the number of fishers are concluded to be to some extent a reflection of variations in the productivity of the ecosystems. Lake Chilwa again is the clearest example of how the size of the fishing population fluctuates with such changes in productivity according to the lake levels. But also in other cases, one can infer that changes in productivity and catch rates to a certain extent regulate the level of effort.

The prevailing biological approach, based on setting effort levels through stock assessment models, is therefore not a solution if the objective of fisheries management is to secure a livelihood for the population. From a biological point of view, the management of fish stocks has to be pragmatic and adaptive. In such an approach, the first step would be to undertake an analysis of trends and variability in catches, catch rates, fishing effort and water levels. This will provide knowledge on the particular behaviour of a system under various conditions. The second step would be to establish a set of important biological indicators that can form the basis for the provision of knowledge that can be used in the decision-making process by the various agents involved in management of these fisheries. A generalized approach towards devising such indicators has to be based on information on:

System variability. Freshwater lakes and rivers can be classified over a range from pulsed to constant environments. In Chapter 5 we have shown that the 'Relative Lake Level Fluctuation', which is based on the seasonal and annual amplitude in water levels related to the depth of the lake, is a simple index that may help in such a classification. For any particular system, changes in water levels can provide a number of other indices that can be related to changes in stocks. For instance, persistence of conditions could relate to short-term, decadal trends in stocks and species composition, while peaks in lake levels could be related to strong year classes of longer-lived species. Lastly, the size of inundated areas at different lake levels could provide another index and be related to variations in catch rates and species composition.

Susceptibility of species to fishing[31]. Applying this concept will indicate that management becomes less relevant from a biological perspective the more resilient a species is to increases in fishing pressure. 'Resilient' species, such as *tilapias* – dominate in many African freshwater fisheries – and more especially 'highly-resilient' species, such as small-sized species like freshwater herrings and small barbs. Indicators of change relate to type and size of species in the catch and lag phases between changes in catch rates and water levels.

Selectivity and scale of operation of fishing patterns. There is a limited danger in increased diversification of fishing patterns with small operational scales, i.e. when fishermen use methods that catch the ubiquitous bucket of fish per day. Such fisheries come very close to an overall unselective and ecologically sound fishing pattern, highly adaptive to changing conditions. The danger increases with increased gear efficiency, whether this arises from investments in better technology or from more intensive use of existing technology. Examples of such indicators are increase in numbers of gear per unit, increased sizes of individual gears and higher investment levels in gear and vessels.

---

[31] This concept is presented in more detail in Chapter 5 in the section on "Susceptibility of fish stocks and species to fishing under environmental variation".

These classifications and associated indicators are highly relevant in designing management systems for SADC freshwater fisheries. Since the impact of fishing activities on the productivity of the stock has proven to be limited, it is difficult from a biological perspective, to see the need for management in pulsed environments or in fisheries targeting resilient or highly-resilient species with a multitude of small-scale individually selective, but overall unselective gears. As long as the increase in fishing effort is only driven by population growth, fish resources in the four lakes (with few but important exceptions) are unlikely to be overexploited. The fisheries will in other words, remain largely biologically sustainable both with and without detailed management measures based on stock assessment models. A truly multispecies/multi-gear fishery cannot be managed on the basis of single-species models alone, though important species for which much information can be obtained could act as important indicators of change, if their changes in biomass or length structure reflect changes pertaining to the whole fishery. Besides this, a multispecies/multi-gear fishery can only be managed on the basis of the aggregations of species in the catch. This means that instead of focusing on individual species, management should be directed to maintaining the integrity of fish communities and ecosystems, including the source of their variation.

For more than a hundred years, fisheries management in SADC freshwater lakes has primarily been about establishing operational rules for the fisheries. This subject has been dealt with in detail in Chapter 3. Here we would only repeat that already in 1952, C.F. Hickling, a fisheries biologist in the Colonial Office in London, considered the effectiveness of most of the restrictions and prohibitions that had been in place in the colonies for almost 90 years as doubtful. He pointed out that licensing of gear or nets requires a large and expensive enforcement staff. Other measures, such as closed seasons, mesh-size regulation and size of fish regulations were also questioned as general management instruments. Nevertheless, despite his reservations towards most of the applied regulations, they have continued to form the basis for management of freshwater fisheries in sub-Saharan Africa, even though variations between countries exist. Our findings generally support Hickling's critique, but, although the relevance of these measures may be debated, they might be both necessary and useful in specific situations in a more adaptive management context. From a biological point of view, what to manage in systems that are adapted to high environmental variability can be summarized in a small number of general rules on an ecosystem level. Specific measures either derive from these rules or relate to specific objectives in order to:

- Maintain the integrity of the ecosystems including their natural, environmentally driven, variation;
- Maintain variability in habitats and avoid habitat destruction;
- Protect important refuges in periods of adverse conditions, for instance through temporary closures.
- There is no need for detailed gear regulations, as long as there is no major increase in efficiency of particular gears. Specific management measures can, nevertheless, be relevant for specific reasons.

Mesh size and other gear regulations. As shown in Figure 5.5, the regulation of gears, including mesh sizes can be interventions that have highly different consequences for different trophic levels within an ecosystem. For instance, employing smaller mesh sizes could increase catch rates over the short term, but at the same time the fishery will become more vulnerable to environmental changes, because the variation in CpUE is larger for smaller-sized species. Large mesh sizes could be needed to protect larger species like the Tigerfish (*Hydrocynus vittatus*) in Lake Kariba, but they are also measures to avoid the fishing of undersized specimens of larger species (i.e. the fish for the future). Choice of mesh size thus depends on the purpose of the fishery. Nevertheless, from a management perspective it is difficult (indeed almost impossible) to redirect the fishery "back again" towards larger fish as it requires very restrictive management measures (e.g. a major increase in the mesh size employed or closing the fishery for a period of time). Measuring its success will take time, at least in the order of a few years to a decade (Pet, 1995; Pet et al., 1996; Densen, 2001). Because of the difficulty in showing the biological impact of applying more restrictive management measures, such measures will most likely undermine the support of fishers where they are imposed.

Closed or protected areas are management measures that can be applied to protect nursery grounds, spawning aggregations or avoid habitat destruction in certain areas. In relation to Lake Malombe, some biologists believe that habitat destruction – in this case the disappearance of submerged weed beds, due either to reduced water levels or to seining – explain why the Chambo (*Oreochromis* spp.) stocks have collapsed. In the case of Lake Malombe, protected areas could therefore perhaps be effective in the recovery of this species by protecting spawning grounds. In Lake Chilwa, it is important to protect pools with fish remaining when the lake has dried up. On the Zimbabwean side of Lake Kariba, closed areas are used to protect spawning areas and nursery grounds, despite the fact that there is no strong biological evidence to support this management measure (Jul-Larsen et al., 1997). In some cases, these management measures are introduced to pursue political objectives by excluding fishers from certain areas in order to protect the interests of more powerful actors (e.g. tourist operators). For instance, in Lake Kariba closed areas have been used to maintain social control over the fishers and actually been used as an instrument to move fishers from certain areas (Malasha, forthcoming). On the other hand, the establishment of a closed area in Lake Mweru was aimed to protect the stocks, although little evidence exists for its effectiveness. But even if these examples may indicate something else, closed areas or no-take areas can, as already mentioned, be useful both in an adaptive management context during periods of adverse conditions, or as a general precautionary measure.

The implementation of closed seasons is a general measure to protect fish during breeding periods. This instrument has been used in Lake Malombe, but has been difficult to enforce. The major effect of a three-month closure in Lake Mweru has been that trade of fish more or less comes to a halt, which reduces general effort levels considerably. Although being heavily contested, the measure is generally considered acceptable by fishermen because they are still able to fish for subsistence and because they then direct their labour towards agriculture (Gordon, 2000; Aarnink, 1997).

Licences are promoted based on a biological rationale. However, in most cases governments enforce licensing mainly to obtain political and social objectives. Hickling also argued that the introduction of licences was mainly about supporting the Native Authority. In line with legislation supporting traditional leaders, e.g. the Administration Act and the Indirect Rule Act[32], fishing licences became a means to generate revenue for this Authority. More than to control fishing effort, licensing remains an important tool for many local chiefs or local authorities to obtain their share of the income generated from fishing activities (Malasha, forthcoming). The biological rationale for a licensing system in small- and medium-sized freshwater lakes in SADC is thus hardly convincing.

---

[32] The acts may have had different names in Zambia and Zimbabwe, and Malawi, but the approach of the British government was almost identical for the three countries, at the time of the Federation of Rhodesia and Nyasaland.

Malasha's research demonstrates that management of fisheries is not so much about managing fish stocks as about managing or controlling fishers and other users of the resource. According to our findings, freshwater fisheries in the SADC region will remain biologically sustainable without any management intervention initiated by governments under the present macro-economic situation and local institutional conditions. The fisheries in small- and medium-sized freshwater lakes in southern Africa can in many ways be considered to be self-managed from a biological perspective. This does not mean that there is no need for management. From a precautionary viewpoint – which we firmly support – there is a need to monitor the development of catches, catch rates, fishing effort, water levels and the maintenance of the integrity of habitats. The challenge will be to use existing data more effectively and to set up effective data collection and monitoring systems based upon longer time series. Biological management may also be required if one wants to promote particular types of fisheries (e.g. angling for tourism). In any case, management will have to be something different from the traditional approach taken in the SADC region so far.

6.3   Social findings and management implications

As shown in Chapter 3, the aggregate fishing effort in SADC freshwaters has increased considerably in the last 50 years and conventional models relate the increase to general social factors like demographic growth and increased demand for fish. However, the lakes under study demonstrate great internal differences in their effort development. For example in Lake Mweru, the number of fishers has steadily increased during the last four decades. In Lake Kariba, the number of inshore fishers has fluctuated a lot and effort is probably not much higher today than it was when the lake was filled up. The number of fishermen increased in Lake Malombe during the 1970s, but already in the 1980s, it started to stabilize and lately it has decreased (Hara, 2000; Donda, 2000). In Lake Chilwa, major fluctuations in the number of fishers have been observed, mainly due to the major water recessions during which the lake dries up completely. During major recessions, fishing operations have been suspended for two or three years.

Only when we break down the concept of fishing effort, is it possible to discover regularized patterns in the development of these fisheries. It proves that most changes in fishing effort are population-driven. Most of the fisheries seem able to absorb great numbers of newcomers who mainly continue to fish in the same manner and with the same gear as their predecessors. Malombe is the only lake where a major investment-driven growth in effort can be observed. The expansion of the Mpumbu fishery in Lake Mweru in the late 1940s is another example of investment-driven growth, but this fishery stopped in the 1950s due to the collapse of Mpumbu. In all other cases, it is population-driven changes we notice and the fisheries are characterized by extensive rates of both in- and out-migration. New people seem not only to enter, but also to go back to other occupations when they consider it worthwhile. Fisheries in southern African freshwaters therefore do not function as a 'last resort' in the sense that once someone has entered there is no way to return to other occupations. It remains a fact, however, that far more people have entered the fisheries over the past 50 years than have left them.

Demographic growth and/or better markets for fish can explain neither the variations between the lakes, nor the internal fluctuations in fishing effort. Even if we accept that these factors play a role, we need to look for additional factors to explain effort development in SADC freshwaters. Fluctuations in the number of fishers are to some extent a reflection of the variations in the productivity of the ecosystems. Lake Chilwa is the clearest example of how the size of the fishing population fluctuates according to lake levels, but also on Kariba we saw how effort fell when the productivity decreased. The fluctuations also reflect changes in the conditions and job opportunities in other sectors of the economy. Improved conditions and better job opportunities elsewhere will immediately result in reduced fishing effort and vice versa. In order to understand the dynamics of freshwater fisheries in the SADC region, they must be placed in a much larger macro-economic context. Only when we see an improvement in the general macro-economic conditions will we be fully able to judge the relative importance of the role of external opportunities.

Furthermore, the local access-regulating mechanisms, which exist everywhere, prove to be important. As argued by Brox (1990), rather than being a management means as such, local access-regulating mechanisms more often reflect how local class and power relations mediate and restrict people's access to resources. In many cases, such mechanisms are not made relevant or they are not effective, but in the cases of Malombe and Kariba we saw how periods of stabilization and reduction can only be understood with reference to such mechanisms.

Finally, we have observed that the use of capital-intensive fishing methods sometimes prevents population-driven increase in effort: the cost of equipment becomes so high that only a few fishers can afford it. This observation is supported by experiences from other parts of the world, such as in the pelagic fisheries in the North Sea employing Dutch, Danish, Scottish and Norwegian vessels, or the off-shore cod fishery out of Iceland or Norway using freezer trawlers, where a reduction in nominal effort (number of fishers or boats) often follows a major change in the technology applied or in the organization within the fishery. The "Lords of Malombe" (Hara and Jul-Larsen, 2003) could remain Lords because so few could cope with the investment costs and just as in the tragedy of the Norwegian Herring fishery (Brox, 1990), the tragedy in Malombe was not caused by an unlimited access of the many, but by the competition between the selected few who remained in the race.

Investment-driven growth is not as inevitable as projected by Gordon (1954) and Hardin (1968). With the prevailing macro-economic conditions in the region, with very little growth and limited funds for investment, it is unlikely that freshwater fisheries will attract investment from the outside, at least in the short- and medium-term perspective. Furthermore, reduced possibilities in labour migration and a complicated, unclear and ambiguous institutional landscape at the local level seem to effectively prevent any investment-driven growth from within the fishing communities. Probably the only case at present where investment-driven growth dominates development is on Lake Victoria.

The geographical and occupational mobility characteristic of fisheries in southern African freshwaters, and the flexibility of the ecosystems is a key to understanding their importance for the countries, as well as for the people involved. They serve as buffers in the national and regional economy and they serve as an important economic safety mechanism for thousands of people. Few cases we know about better illustrate Brox's (1990) argument that the problem with common property theory is not what it says, but what it omits to say. There is no doubt that the commons can lead to ecological tragedies; we have seen it in the case of the Mpumbu fishery in Mweru and the Chambo fishery in Malombe, but the theory omits to tell us that the commons also may play a very positive role, both for the individuals as well as for society. We will argue that it is important to maintain the southern African freshwater fisheries as commons – as valuable local and even national assets through their function as buffers and safety valves – in particular if poverty reduction is taken into consideration. Instead of enforcing severe restrictions on resource use, the freshwater fisheries should continue to be used as a safety valve for local people as well as migrants – facilitating their adaptation to changes in the macro-economic environment and in their occupational surroundings. In order to do so, it is important that management can control, or more correctly prevent, large investment in the fisheries. If investment-driven growth is allowed to take place, the consequences may become serious. In such circumstances, well considered and timely interventions are required from decision-makers in order to avoid building-up fishing power.

Here it is important to distinguish between investments by people from outside the fishing communities and investments made by people from inside. Our research indicates that a combination of government prohibitions and local access-regulating activities are quite effective in limiting massive investment-driven growth from the outside. In such cases, governments may intervene and thereby protect local interests. This was in fact the situation in Lake Mweru some years ago, where trawling was forbidden in order to protect local fisheries and food security.

Paradoxically, government intervention is often more problematic if investments come from inside the communities; neither government nor the local communities seem particularly able to cope with such a situation. It is extremely difficult for the local institutions to prevent insiders from investing in the fisheries, because insiders who are in a position to do so are usually powerful people and therefore in a strong position in the decision-making processes. Government intervention to avoid an investment-driven increase of effort is thus likely to be unsuccessful because of lack of local support. However, we see that investments from within the community often require financial resources from labour migration or other external sources, as well as an institutional landscape very different from what is generally found in rural southern Africa. The case of Malombe also reminds us that investors tend to diversify their activities. As a consequence, they consider themselves as much as farmers and/or businessmen than as fishermen. As long as the macro-economic conditions do not improve, we do not consider investment from inside the fishing communities to be a major management problem. In a longer term perspective, or where investment-driven growth already dominates the development, cooperation between government and the fishing communities appears to be the only way to improve the situation.

From a social and economic perspective, it is important to emphasize that population-driven growth in effort also reduces catch rates, even if by far less than investment-driven increases. Reduced catch per unit of effort (CpUE) certainly has negative economic consequences for the individual fishers. There is therefore a balance to strike between the need for individual incomes on the one hand, and the need for a buffer in the economy and a safety valve on the other. We are not certain, however, that government can strike such a balance in an effective way, but there is a good chance that the local access-regulating mechanisms in most cases can deal with it. The higher the population pressure becomes, the more we must assume that local access-regulating mechanisms are put into function. We saw in Box. 4.1 how they did in Kariba when the pressure on the land resources along the shore became too high.

Given that such mechanisms are deeply embedded in the local institutional landscape, there are few possibilities for government to play an effective role. But because population-driven growth in effort leads to an intensification in the level of conflicts, government has an important role to play in conflict mediation and resolution. As long as the conflicts are between groups that are economically or politically equal, disputes will probably best be resolved locally. However, in situations where conflicts are less embedded in local conditions, or where the lines of conflict follow clear differences in economic or political status and become more permanent (e.g. as in the Mweru fishery on the Congolese side, where the great trader Katebe Katoto had a lot of control), there is – ideally speaking – also a need for active support by governments of the weakest party in the conflicts. This is also the case if conflicts are between actors making alternative use of freshwater resources (e.g. fishers vs. tourist operators in Lake Kariba).

This type of social and/or economic management requires sets of information and data that at present are virtually non-existent. It is interesting that despite the change in rhetoric about the need for economic and social considerations in fisheries management, this has had very little impact on how the authorities organize their fisheries services, how the service is staffed or what sort of data they collect. To manage the fisheries in the manner just outlined will require the elaboration of some economic (e.g. cost-earning and investment) and social indicators (e.g. dependency of fishing and migration) and their integration in the data collection and analysis systems.

6.4 Management, co-management or no management in southern African freshwater fisheries?

We started out in this report by referring to some uncomfortable feelings that we had related to the debate between management and co-management approaches. Before we end, we would like to explicitly address the question: to what extent can the management and the co-management approaches be useful for the governance of southern African freshwater fisheries? What kind of role should governments have vis-à-vis local communities, or should the fisheries be left to manage themselves? Throughout the report, the findings demonstrate that the answers to these questions are somewhat more complicated than what the first reaction may indicate. Even if the fisheries under study in many ways can be seen as self-regulatory with little need for external intervention, we have also shown that there is a need to monitor their development, and that management measures soon may be needed if major investment-driven growth in effort should take place. Changes in existing investment levels and patterns may change the situation dramatically in a limited period of time and lead to serious bio-ecological problems. This shows how the management approach and common property theory on which it relies must still be considered relevant. Even limited investment-driven changes may easily lead to tragedy scenarios as we have already shown in some limited cases. However, despite the lack of effective resource management regimes, investment-driven growth in fishing effort has not been characteristic of the fisheries in southern African freshwaters. As has been the situation over the past 50 years, nothing indicates that a 'no management' solution on the part of the governments will lead to biological tragedies in these commons. On the contrary, too much government intervention may result in the fisheries losing some of their important social and economic functions as buffers and safety valves for a great number of people with limited possibilities in periods of stress. In that respect, too much reliance on CPT is of little use and may prevent us from capturing the present strengths of the freshwater commons.

The centralized scientific management approach often taken by the various SADC governments has been inappropriate for a number of reasons. But in order to understand why they have chosen such strategies, it is impossible not to emphasize the role of colonialism. Many of the fisheries regulations can in practice still be traced back to the colonial administrations from the first part of last century, (Hara, 2000; Malasha, forthcoming.). For Zambia and Zimbabwe, Malasha concludes that: "despite the numerous amendments attempted by the two countries, the principal objectives of fisheries management have remained the same since the advent of colonialism about a hundred years ago." After political independence, the centralized approach was maintained for at least two reasons. First, it was an approach that had been applied by the industrialized countries for a long time. The people in the new administrations had all received their training and professional experience from colonial institutions and staff who were convinced that their way of conceptualizing the relationship between man and nature was the only 'natural' one. Secondly, the new administrations had also learned another lesson from those previously in control: under the pretext of concern for natural resources, the centralized management approach had proven to be a well-suited policy to control people in rural areas. The different political regimes that succeeded colonial rule, whether they called themselves socialist as in Zambia and Zimbabwe, or autocratic, which was the situation in Malawi for many years, therefore had good reasons not to give up this policy. The history of one hundred years of resource management in the region reveals that government, and its individual representatives, often act on the basis of their own interests rather than from a consideration of what type of regulation is appropriate.

Our report also emphasizes how increased population-driven growth in effort has already increased the level of conflicts between different groups of users and that this often requires monitoring as well as active involvement from government. This type of fisheries management can certainly not be solved by a centralized approach and can only be effective with an active participation and influence from the people directly involved. The neo-institutional framework of analysis that inspires the co-management approach has some great advantages compared to CPT. We have shown in Chapter 4 how a neo-institutional framework may be of help in understanding why some of the assumptions in CPT do not apply in the southern African fisheries. But its conception of institutions is somewhat too simplistic to fit the realities of the African rural communities. Only by taking empirical analyses of the existing institutional landscape into account, does the framework become more useful as a guide for concrete action. CPT is a conceptual model and can only be useful for designing management measures if it is integrated into an empirical analysis.

Most people involved will agree that a uniform concept of co-management does not exist and that co-management requires a reorganization of the institutional set-up. However, this is not a question of 'constructing' local institutions that can cooperate with central authorities in managing the fisheries. Such an approach was already tried in what the colonial powers labelled "indirect rule". The term "community-based co-management" has been suggested in order to get some meaning into the concept (Holm *et al.*, 2000; Jentoft, 2000). In this approach, fishing communities are considered the platform to ensure local participation. There are some serious problems connected to this concept, since our study shows that "pure" fishing communities hardly exist along the lakes in southern Africa. Most communities have a very diverse occupational structure, and fishing is only one among several occupations that provide the livelihood for the local people. In addition, many fishers have two homes – one at the lakeshore and another somewhere else. It is therefore almost impossible to create boundaries between fishers and non-fishers with regards to the representation of "the community" in co-management arrangements.

Furthermore, according to the results and lines of thought in the research of Berry (1989, 1993) and others, the main reason for local actors to get involved in government-initiated projects is to secure and improve their access to resources. Granting access rights to specific groups is therefore problematic. Who should be granted such rights? Who is a "genuine" fisher? Our study shows that everybody is a (potential) fisher, and therefore that no one should be excluded from the decision-making process. This complicates a management system based entirely on the voice of the (existing) fishing communities because it will not reflect the concern of keeping the fisheries as safety valves in people's strategies towards changing (and until now deteriorating) macro-economic conditions. If co-management primarily becomes a tool for allocation of access rights to local communities, it will prevent the freshwater lakes from remaining commons and safety valves. Since a policy including all potential users will be impossible, the only solution to this dilemma is that government takes upon itself to assure that the collective concerns are voiced and taken into account. We are not arguing that co-management cannot be an appropriate approach at all. Our concern is rather the importance of distinguishing between the allocation of access rights, the making of operational rules to control fishing activities and active participation in conflict resolution. In relation to the last two aspects, co-management can arguably be a useful tool to manage fisheries in the region (Normann *et al.*, 1998).

In most cases where co-management initiatives have been launched, they have coincided with a change of political regime towards more democratic principles of rule, and with the initiation of Structural Adjustment Programmes on request of the World Bank and supported by most of the international donor agencies. This process has led to the initiation of decentralization policies in order to reduce public spending and the assumption that it will bring more local influence in the decision-making processes. Many studies of the Structural Adjustment Programmes (e.g. Mosly *et al.*, 1991) have shown how quickly recipient governments find strategies to by-pass the conditions imposed upon them by the donors. Studies on decentralization processes (e.g. Bierschenk and Olivier de Sardan, 1998) demonstrate how the main result of political decentralization is often a shift of influence and decision-making from one elite to another.

Also, the experiences of setting up co-management in the region have not been very encouraging up till now. Most arrangements have tended to exclude user groups from the decision-making process and from influencing who should participate in the making of operational rules for the fisheries (Sverdrup-Jensen and Raakjaer Nielsen, 1998). Generally speaking, management of freshwater fisheries is still very much in control of governments, and the negotiating position of user groups versus that of governments is consequently comparatively low. As Chirwa (1997:69) points out: "The FD's [Fisheries Department's] position of patronage means that the local user communities are the recipients rather than the initiators of decisions. They, themselves, are managed, together with their resources, by the Fisheries Department." This statement seems to be applicable to most of the examples of co-management within the region. Malasha's comparative study (forthcoming) on the introduction of co-management in Zambia and Zimbabwe (Lake Kariba) demonstrates that, even if the co-management processes developed very differently in the two countries, control of people seems to have been a more important concern than considerations for the resources. The only difference was that it was the desire for control by the chiefs and the Kapenta operators which became important in Zambia, while it was the wishes of government which have dominated in Zimbabwe. It is probably correct to say that there is little difference between much of today's co-management arrangements and those of "indirect rule" that the British colonial power established a century ago.

In Chapter 5, we said that a major challenge in the management of small- and medium-sized freshwater lakes in SADC countries is to improve the knowledge base, and we argue that participation of the fishing communities is a precondition for implementing the monitoring systems that we advocate. It is simply not possible to move away from the classical management paradigm towards a more pragmatic and adaptive approach to management without involving the direct users in the process. But instead of opting for advanced 'social engineering' – favoured by many donor agencies and non-governmental organizations – we propose a more modest strategy. Co-management must imply a process of mutual adaptation, where one tries to establish a convergence between government policies and the local institutional structures. Given the great institutional ambiguity and lack of clarity, which we have shown to exist at the local level, even this is by no means an easy task, but rather a long-term process with a lot of "muddling through". It will have to be a learning process, and unfortunately, the specific design has to be tailor-made to suit the needs in each specific fishery or lake.

In this respect, it is a problem that government departments and their fisheries research extension services have not been reorganized, and the inputs to be accommodated by government have not been changed as a result of the changes in emphasis towards a more social and economic fisheries management (Donda, 2000). Co-management is mainly an arrangement to ensure communication between governments and communities. From a resource management perspective, it is important to establish a dialogue on the necessity of various management measures, as well as how to measure their impact. It involves an attempt at grass root level to develop new indicators for the development of the fisheries, which can be accepted by government as well as by the communities.

Irrespective of the political systems experienced in the past, fishers' trust in government authorities has always been (at best) moderate. They have hardly ever found themselves at the winning end of the relationship and therefore, initiatives to establish co-management taken by government authorities are likely to be met with profound scepticism. As we just said, this is not without good reason; fishermen are suspicious about the motives and the sincerity of government authorities when they propose collaboration and the sharing of responsibilities. In their experience, the government always sets the rules and regulations, and co-management has entered a trajectory where fishers are only involved in the enforcement of the operational rules set by the government. The democratization discourse may have given fishers and other local actors an incentive to give collaborative management arrangements a try. But, if co-management continues on its present path, excluding stakeholders from defining management objectives or general support from government agencies, it may be pertinent to ask what benefit they can have from co-management arrangements.

To conclude, we support Brox's argument (1990) that common property theory still may have its strength as a conceptual model, but that it is definitely not a management tool. Government's active participation in management is certainly needed to avoid major investment-driven growth of fishing effort and to ensure that the small- and medium-sized freshwater lakes can remain a safety valve or a buffer for people in their struggle against macro-economic deterioration. In this way, the fisheries can continue to serve as important means to alleviate poverty and redistribute economic resources. We are aware of the fact that this tends to contradict mainstream management approaches that focus on limiting people's access to the fisheries. As argued in this report, it is however, clear that this approach falls short for small- and medium-sized freshwater lakes in southern Africa.

But governments' involvement has also to be tuned to voices and needs at the local level and co-management can certainly represent an option as long as today's approach is changed. Co-management is fine as long as it focuses on the "real" or "appropriate" issues (i.e. conflict resolution and avoiding investment-driven growth of effort) and it is seen as a process of empowering communities and a means to integrate the regulation of fisheries in the general development of the communities. However, if it only continues to be seen through the lens of biological conservation, co-management is not really required in these systems characterized by high system variability and with fisheries targeting species highly resilient to fishing. From a precautionary perspective, there is a need to monitor the fishery and to communicate/evaluate the results to the various users.

7. REFERENCES

The references do not follow the FAO format. They are inserted as presented by the author.

Aarnink, B.H.M. (1997). *Fish and cassava are equally important in livelihood strategies of women in Mweru-Luapula Fishery*, DoF/ML/1997/Report no. 40. Nchelenge, Zambia, Department of Fisheries.

Abila, R. & E. G. Jansen (1997), *From local to global markets; the fish exporting and fishmeal Industries of Lake Victoria – Structure, strategies and socio-economic impacts in Kenya*, Report no.2, Socio-economics of Lake Victoria, Nairobi, IUCN.

Acheson J.M., (1981), Anthropology of fishing, *Annual Review of Anthropology*, 10: 275-316.

Adams P.B., (1980), Life History Patterns in Marine Fishes and Their Consequences for Fisheries Management, *Fish. Bull.* 78: 1-12.

Adams W.M., (1992), *Wasting the rain: rivers, people and planning in Africa*, London, Earthscan.

Allen K.R., (1971), Relation between production and biomass. *J. Fish. Res. Board Can.* 28: 1573-1581.

Allison E.H., Patterson G. Irvine, K. Thompson A.B. & A.Menz, (1995). The pelagic ecosystem, in: *The fishery potential and productivity of the pelagic zone of Lake Malawi/Nyassa*, Menz A. (ed.), Kent, Natural Resources Institute, 351-368.

Andersen K.P. & E. Ursin, (1977), A multispecies extension to the Beverton and Holt theory of fishing, with accounts of phosphorous circulation and primary production. *Medd. Dan. Fisk. Havunders.* 7: 319-435.

Anderson D. & R. Grove (eds.), (1987), *Conservation in Africa; people policies and practice*, Cambridge University Press, Cambridge.

Andrewartha H.G. & Birch, L.C. (1954) *The Distribution and Abundance of Animals*, Chicago, University of Chicago Press.

Anonymous (1992), Working group on assessment of Kapenta (Limnothrissa miodon) in Lake Kariba (Zambia and Zimbabwe). Zambia/Zimbabwe SADC Fisheries Project Report No 11. Kariba 4-17/3 1992. L.K.F.R.I., P.O. Box 75, Kariba, Zimbabwe.

Bâcle J. & Cecil, R., (1989). La pêche artisanale en Afrique Sub-Saharienne. Sondages et recherches, Hull, Agence Canadienne de Dévelopment International.

Bailey K.M. & E.D.Houde, (1989) Predation on early developmental stages of marine fishes and the recruitment problem. *Adv. Mar. Biol.* 25: 1-83.

Balon E.K., (1984), Patterns in the Evolution of Reproductive Styles in *Fishes, in Fish Reproduction – Strategies and Tactics*, G.W. Potts & R.J. Wootton (eds), London, Academic Press: 35-53.

Bayley P.B., (1988), Accounting for effort when comparing tropical fisheries in lakes, river-floodplains and lagoons, *Limnol. Oceanogr.*, 33(4): 963-972.

Beddington J.R., (1984), The Response of Multispecies Systems to Perturbations, in *Exploitation of Marine Communities,* R.M. May (ed.), Dahlem Konferenzen, Life Sciences Research Report 32, Berlin, Springer-Verlag: 209-225.

Beddington J.R. & May R.M., (1982), The Harvesting of Interacting Species in a Natural Ecosystem, *Scien. Am.* 247(5): 42-49.

Begon M., Harper J.L., Townsend C.R., (1996), *Ecology. Individuals, populations and communities.* 3rd edition, Oxford, Blackwell Science Ltd.

Beinart W., (ed.), (1989), Special issue on the politics of conservation in Southern Africa, *Journal of Southern African Studies*, 15 (2).

Bennett B.A., (1993), The fishery for white steenbras Litognathus lithognatus off the Cape coast, South Africa, with some considerations for its management. *S. Afr. J. Mar. Sci.* 13: 1-14.

Berkes F., Folke C. and Colding J. (eds.), (1998), *Linking social and ecological systems: management practices and social mechanisms for building resilience*, Cambridge, Cambridge University Press.

Bernacsek G.M., (1989), Improving Fisheries Development Projects in Africa, in Proceedings on the World Symposium on Fishing Gear and Fishing Vessel Design. St. John's New Foundland, Marine Institute: 155-164.

Berry S., (1985*), Fathers Work for their Sons*, Berkley, University of California Press.

Berry, S., (1989), Social institutions and access to resources, *Africa* 59: 41-55.

Berry, S., (1993), *No condition is permanent: The social dynamics of agrarian change in sub-Saharan Africa*, Madison: University of Wisconsin Press.

Berry S., (1997), Tomatoes, land and hearsay: Property and history in Asante in the time of structural adjustment, *World Development* 25: 1225-1241.

Berry S., (2001), *Chiefs know their boundaries; essays on property, power and the past in Asante, 1896-1996*, Portsmouth Oxford and Cape Town, Heinemann, James Currey & David Philip.

Beverton R.J.H., (1990), Small marine pelagic fish and the threat of fishing; are they endangered. *Journal of Fish Biology*, 37 (suppl.A): 5-16.

Beverton R.J.H., & Holt S.J., (1957), On the Dynamics of Exploited Fish Populations. Fishery Investigations, Minist. Agric. Fish. Food, Ser. II.

Bierschenk T. & J-P. Olivier de Sardan, (1998), *Les pouvoir au village; Le Bénin rural entre démocratisation et décentralisation,* Paris, Karthala.

Bjorstad O.N., Grenfell B.T., (2001), Noisy clockwork: time series analysis of population fluctuations in animals, *Science* 293: 638-643.

Bos A.R., (1995), A limnological survey on Lake Mweru, Zambia. DOF/ML/1995/ 27, Nchelenge, Zambia, Department of Fisheries.

Bossche J.P. van den & Bernacsek G.M., (1990), *Source book for the inland fishery resources of Africa*, Vol. 1 and 2, Fisheries Technical Paper No.18 (1-2), Rome, FAO.

Boudreau P.R. & L.M. Dickie, (1992), Biomass spectra of aquatic ecosystems in relation to fisheries yield. *Canadian Journal of Fisheries and Aquatic Science* 49: 1528-1538.

Bowen S.H., (1988), Detritivory and Herbivory, in Biologie et Ecologie des Poissons d'eau Douce Africains - Biology and Ecology of African Freshwater Fishes, Lévêque, C., Bruton, M. & Ssentongo, G. (eds.), Collection Travaux et Documents No. 216, Paris, ORSTOM: 243-247.

Boyce M.S., (1984), Restitution of r- and K-selection as a Model of Density-dependent Natural Selection. *Ann. Rev. Ecol. Syst.* 15, 427-447.

Brandt, A. von, (1984), *Fish Catching Methods of the World*, (3rd ed.), Oxford, Fishing News Books Ltd.

Brox O., (1990), The common property theory: epistomological status and analytical utility. *Human Organization* 49(3): 227 - 235

Caddy J.F., (1991), Death rates and time intervals: is there an alternative to the constant natural mortality axiom? *Rev. Fish Biol. and Fisheries* 1: 109-138.

Caddy, J.F. & Csirke J., (1983), Approximations to sustainable yields for exploited and unexploited stocks, *Océangr. trop.* 18: 3-15.

Caddy J.F., Mahon R., (1995), *Reference points for fisheries management*, Fisheries Technical Paper 347, Rome, FAO.

Carpenter S.R., Kitchell J.F. & Hodgson J.R., (1985), Cascading trophic interactions and lake productivity, *Bioscience* 35: 643-639.

Carruthers J., (1995), *The Kruger National Park; a social and political history*, Pietermaritzburg, University of Natal Press.

Chaboud C., Charles-Dominique E. (1991), Les pêches artisanales en Afrique de l'Ouest : état des connaissances et évolution de la recherche, in *La recherche face à la pêche artisanale*, Durand J.R., Lemoalle J. et Weber J. (eds.), Symposium International ORSTOM-IFREMER, (France), 3-7 juillet 1989 Montpellier, Paris, ORSTOM: 99-143.

Chanda B., (1998), Effects of weir fishing on commercial fish stocks of the Bangweulu swamp fisheries, Luapula province, Northern Zambia, M phil. thesis, Bergen, Department of Fisheries and marine Biology, University of Bergen.

Charles A.T., (2001), *Sustainable fishery systems*, Oxford, Blackwell Science Ltd.

Charnov E.L. & Schaffer W.M., (1973), Life-history Consequences of Natural Selection: Cole's Results Revisited, *American Naturalist* 107: 791-793.

Chauveau J-P. & E. Jul-Larsen, (2000), Du paradigme halieutique à l'anthropologie des dynamiques institutionelles, Introduction in *Les pêches piroguières en Afrique de l'Ouest;pouvoirs-mobilités-marchés*, J-P.Chauveau, E. Jul-Larsen & C. Chaboud (eds.), Paris, Karthala: 9-85.

Chipungu, P. & H. Moinuddin, (1994), Management of the Lake Kariba fisheries, Kariba, Zambia-Zimbabwe SADC fisheries project, project report no.32.

Chirwa, W. C., (1996), Fishing rights, ecology and conservation along southern Lake Malawi, 1920-1964, *African Affairs*, 95: 351-377.

Chirwa, W. C., (1997), The Lake Malombe and Upper Shire River fisheries co-management programme: An assessment, in Normann *et al.* 1998.

Chitembure, R.M., Songore, N. & Moyo, A., (1999), 1998: Joint Fisheries Statistical Report, Kariba, Lake Kariba, Zambia-Zimbabwe SADC Fisheries Project Report No. 59.

Christensen, M.S., (1993), The artisanal fishery of the Mahakam River floodplain in east Kalimantan, Indonesia. III. Actual and estimated yields, their relationship to water levels and management options, *Journal of Applied Ichthyology* 9: 202-209.

Clark B.M., Bennett B.A. & Lamberth S.J., (1994a), A comparison of the ichthyofauna of two estuaries and their adjacent surf zones, with assessment of the effects of beach seining on the nursery functions of estuaries for fish. Africa, *S. Afr. J. Mar. Sci.* 14: 121-131.

Clark B.M., Bennett B.A. & Lamberth, S.J., (1994b), Assessment of the impact of commercial beach-seine netting on juvenile teleost populations in the surf zone of False bay, South Africa, *S. Afr. J. Mar. Sci.* 14: 255-262.

Coates D., (2002), Biodiversity and fisheries management opportunities in the Mekong river basin, in: GEF/UNEP Conference on Fisheries and sustaining biodiversity in the New Millenium Workshop Blue Millenium, Managing Global Fisheries for Biodiversity, June 25-27 2001, Victoria, Canada, World Fisheries Trust City.

Coenen E.J., (1994), Frame survey results for Lake Tanganyika, Burundi and comparisons with past surveys, GCP/RAF/271/FIN-TD/18, Bujumbura, FAO.

Colson E., (1971), *The social consequences of resettlement; the impact of Kariba resettlement upon the Gwembe Tonga.* Manchester, Manchester University Press.

Connell J.H., (1978), Diversity in tropical rain forests and coral reefs, *Science* 199: 1302-1310.

Connell J.H. & Sousa W.P., (1983), On the evidence needed to judge ecological stability or persistence, *American Naturalist* 121:789-824.

Corten, A., (2001), Herring and climate; changes in the distribution of North Sea herring due to climate fluctuations, PhD thesis, Groningen, University of Groningen.

Coulter, G.W. (ed.), (1991), *Lake Tanganyika and its life*, London, Oxford University Press.

Cowx I. G. & Kapasa C. K., (1995), Species changes in reservoir fisheries following impoundment: the case of Lake Itezhi-tezhi, Zambia, in *Impact of Species Changes in African Lakes*, T.J. Pitcher & P.J.B. Hart (eds.) London, Chapman & Hall. 321-332.

Crul R.C.M., (1993), Limnology and Hydrology of Lake Victoria, Paris, UNESCO.

Crul R.C.M., (1995a), Conservation and Management of the African Great Lakes, Victoria, Tanganyika and Malawi. Studies and reports in hydrology UNESCO/IHP-IV Project M-5.1, Nijmegen, Crul Consultancy/UNESCO.

Crul, R.C.M., (1995b). Limnology and Hydrology of Lake Tanganyika. Paris, UNESCO.

Cushing D.H., (1974), The possible density-dependence of larval mortality and adult mortality in fishes, in *The Early Life History of Fish*, J.H.S. Blaxter (ed.), Berlin, Springer Verlag: 103-111.

Cyr H., Downing J.A. & Peters R.H., (1997), Density-body size relationships in local aquatic communities, *Oikos* 79: 333 - 346.

Darwin C., (1972) [1859], *On the Origin of Species by Means of Natural Selection*, American Library, New York, Mentor.

Densen W.L.T.van, (2001), On the perception of time trends in resource outcome. Its importance in fisheries co-management, agriculture and whaling, PhD thesis, Enschede, Twente University.

Deshmukh I., (1986), *Ecology and Tropical Biology*, Blackwell Scien. Publ.

Dickie L.M., (1972), Food Chains and Fish Production, ICNAF Spec. Publ. 8: 201-219.

Doak D.F., Harding E.K., Marvier M.A., O'Malley R.E. & Thomson D., (1998), The statistical inevitability of stability-diversity relationships in community ecology, *American Naturalist* 151: 264-276.

Donda S., (2000), Theoretical advancement and institutional analysis of fisheries co-management in Malawi. Experiences from lakes Malombe and Chiuta, PhD thesis, Aalborg University.

Egerton F.N., (1973), Changing concepts of the balance of nature, *Quarterly Review of Biology* 48: 322-350.

Ellis J.E. & D.M. Swift, (1988), Stability of African pastoral ecosystems: Alternate paradigms and implications for development, in: *Journal of Range Management*, Vol. 41 (6): 450-459.
Estes J.A., (1979), Exploitation of Marine Mammals - r selection of K strategists? *J. Fish. Res. Board Can.* 36: 1009-1017.

Evans D.W., (1978), Lake Bangweulu: a study of the complex and fishery. Fisheries Service Reports Zambia, Chilanga, Fisheries Research Division, P.O. Box 350100.

Fairhead J. & M. Leach, (1996), *Misreading the African Landscape. Society and ecology in a forest-savanna mosaic*, Cambridge, Cambridge University Press.

FAO, (1992), Review of the state of world fishery resources, Part 1. The marine resources. FAO Fish. Circ. No. 710, Rev. 8, Part 1, Rome, FAO.

FAO, (1993). *Fisheries management in the South-East arm of Lake Malawi, the Upper Shire River and Lake Malombe, with particular reference to the fisheries on chambo (Oreochromis spp.)*. CIFA Technical Paper 21, Rome, FAO.

FAO (1999), *Guidelines for the routine collection of capture fisheries data*. Prepared at the FAO/DANIDA expert consultation. Bangkok, Thailand, 18 – 20 May 1998, Fisheries Technical Paper No 382, Rome, FAO.

FAO, (2000a), FAOSTAT: http//apps.fao.org/fishery/fprod1-e.htm

FAO, (2000b), FAOSTAT: http://apps.fao.org/lim500/wrap.pl?Population.LTI&Domain

Fay, C., (1994), Organisation sociale et culturelle de la production de pêche: morphologie et grandes mutations, in J.Quensière (ed.): 191-208.

Fenchel T., (1977), *Almen Økologi*, Copenhagen, Akademisk Forlag.

Ferguson J., (1999), *Expectations of Modernity. Myths and meanings of Urban Life on the Zambian Copperbelt*, Berkeley, University of California Press.

Fox W.W., (1970), An exponential yield model for optimizing exploited fish populations, *Trans. Am. Fish. Soc.* 99: 80-88.

Furse M.T., Morgan P.R. & Kalk M., (1979), The fisheries of Lake Chilwa, in: Lake Chilwa. *Studies in a tropical ecosystem*, Kalk M., McLachlan A.J. & Williams C.H. (eds.), The Hague, Dr. W. Junk: 175 - 208.

Geertz C., (1963), *Agricultural involution: the process of ecological change in Indonesia*, Berley, University of California Press.

Gobert, B., (1994), Size structures of demersal catches in a multispecies multigear tropical fishery. *Fisheries Research* 19: 87-104.

Godø O.R., (1990), Factors affecting accuracy and precision in abundance estimates of gadoids from scientific surveys. Dr. Philos Thesis. Bergen, University of Bergen.

Gorden S. H., (1954), The economic theory of a common property resource: the fishery, *Journal of Political Economy*, 62: 124-142.

Gordon D.M., (2000), The making of a hinterland. Environment and politics in Mweru-Luapula from the 1880s to the 1990s, PhD thesis, Princeton, Department of History, Princeton University.

Gordon, D.M., (2003), Technological Change and Economies of Scale in the History of Mweru-Luapula's Fishery (Zambia and Democratic Republic of the Congo), in *Management, co-management or no-management? Major dilemmas in southern African freshwater fisheries*, E.

Jul-Larsen, J. Kolding, R. Overå, J. Raakjær Nielsen & P.A.M. van Zwieten, FAO Fisheries Technical Paper 426/2. Rome, FAO.

Graham M., (1935), Modern theory of exploiting a fishery, and application to North Sea trawling. *J.Cons. Perm. Int. Expl. Mer* 10: 264-274.

Greboval D., Bellemans M. & Fryd M., 1994, *Fisheries characteristics of the shared lakes of the East African rift*. CIFA Tech. Pap. No. 24. Rome, FAO.

Growe R., (1988), Conservation and Colonial expansion; a study of the evolution of environmental attitudes and conservation policies on St Helena, Mauritius and in the India 1660-1860, PhD thesis, Cambridge, University of Cambridge.

Gulland J.A., (1971), *The Fish Resources of the Oceans*. West Byfleet, Fishing News Books.

Gulland J.A., (1982), The management of tropical multispecies fisheries, in *Theory and management of Tropical Fisheries*, Pauly D. & Murphy G.I. (eds), ICLARM Conf. Proc. No. 9, Manila, ICLARM: 287-298.

Gulland J.A. & Garcia S., (1984), Observed Patterns in Multispecies Fisheries, in *Exploitation of Marine Communities*, R.M. May (ed.) Dahlem Konferenzen, Life Sciences Research Report, 32, Berlin, Springer Verlag: 155-190.

Gunderson D.R., (1980), Using r-K Selection Theory to Predict Natural Mortality, *Can. J. Fish. Aquat. Sci.* 37: 2266-2271.

Gunderson D.R., & Dygert P.H., (1988), Reproductive effort as a predictor of natural mortality rate, *J. Cons. int. Explor. Mer* 44: 200-209.

Guyer J., (1997), *An African niche economy; farming to feed Ibadan, 1968-88*, Edinburgh, Edinburgh University Press.

Haakonsen J.M. (1992), Industrial vs. artisanal fisheries in West Africa: The lessons to be learnt, in *Fishing for development*, Tvedten I. & B. Hersoug (eds.), Uppsala, Nordic Africa Institute: 33-53.

Hannesson R., (1993), Bioeconomic Analysis of Fisheries, Oxford, FAO/Fishing News Books.

Hara M.M., (2000), Could co-management provide a solution to the problems of artisanal fisheries management on the Southeast arm of lake Malawi, PhD thesis, Cape Town, University of Western Cape.

Hara M.M.; (2001), Could marine resources provide a short-term solution to declining fish supply in SADC inland countries? The case of horse mackerel, *Food policy* 26(1): 11-34.

Hara M.M. & E. Jul-Larsen, (2003), The Lords of Malombe; An analysis of fishery development and changes in effort, in *Management, co-management or no-management? Major dilemmas in southern African freshwater fisheries*, E. Jul-Larsen, J. Kolding, R. Overå, J. Raakjær Nielsen & P.A.M. van Zwieten, FAO Fisheries Technical Paper 426/2. Rome, FAO.

Hardin G., (1960), The competive exclusion principle, *Science* 131:1292-1297.

Hardin G., (1968), The tragedy of the commons, *Science* 162: 1243-1247.

Henderson H.F. & Welcomme, R.L., (1974), *The relationship of yield to morpho-edapic index and numbers of fishermen in African inland fisheries*, CIFA Occas. Pap. No. 1, Rome, FAO.

Hilborn R. & Walters C.J., (1992), *Quantitative Fisheries Stock Assessment – Choice, Dynamics and Uncertainty*, New York, Chapman & Hall.

Hoben A., (1995), Paradigms and politics: the cultural construction of environmental policy in Ethiopia, *World Development* 23 (6): 1007-1022.

Hoenig J.M., (1983), Empirical Use of Longevity Data to Estimate Mortality Rates, Fish. Bull. 82(1): 898-903.

Hoggarth D.D. & Utomo A.D., (1994), The fisheries ecology of the Lubuk Lampam river floodplain in South Sumatra, Indonesia, *Fisheries Research* 20: 191-213.

Hoggarth D.D., Cowan V.J., Halls A.S., Aeron T., M., McGregor J.A., Garaway C.A., Payne A.I. & Welcomme R.L., (1999a), *Management guidelines for Asian floodplain river fisheries, Part 2: Summary of DFID research*, Fisheries Technical Paper No. 384/2, Rome, FAO.

Hoggarth D.D., Cowan V.J., Halls A.S., Aeron T., M., McGregor J.A., Garaway C.A., Payne A.I. & Welcomme R.L., (1999b), *Management guidelines for Asian floodplain river fisheries, Part 1: A spatial, hierarchical and integrated strategy for adaptive co-management*, Fisheries Technical Paper No. 384/1, Rome, FAO.

Holling C.S., (1973), Resilience and stability of ecological systems, *Ann. Rev. Ecol. Syst.* 4: 1-23.

Holm P., B. Hersoug & S.A. Rånes, (2000), Revisiting Lofoten: Co-managing Fish Stocks or Fishing Space? *Human Organisation* 59(3).

Horn H.S., (1974), The ecology of secondary succession, *Ann. Rev. Ecol. Syst.* 5:25-37.

Horn H.S., (1978), Optimal Tactics of Reproduction and Life-History, in *Behavioural Ecology: an evolutionary approach*, J.R. Krebs & N.B. Davies (eds.) Oxford, Blackwell Scientific Publications: 411-429.

Huston M., (1979), A general hypothesis of species diversity, *American Naturalist* 113: 81-101.

Hviding E. & E. Jul-Larsen, (1995), *Community-Based Resource Management in Tropical Fisheries*, Windhoek, University of Namibia.

Hyden G., (1983), *No shortcuts to progress. African development management in perspective*, Berkley & Los Angeles, University of California Press.

ICES, (1988), Report on the multispecies assessment working group. Copenhagen, 1-8 June 1988. Doc. CM 1988/Assess:23 (mimeo).

Jansen E.G., 1997, *Rich fisheries – poor fisherfolk; some preliminary observations about the effects of trade and aid in the Lake Victoria fisheries*, Report no.1, Socio-economics of Lake Victoria, Nairobi, IUCN.

Jentoft S., (2000), The community. A missing link of fisheries management, *Marine Policy* 24.

Jul-Larsen E., (1995), The politics of resource management; an assessment of the sociological aspects in connection to developing resource management regimes on Lake Kariba. Kariba, Zambia-Zimbabwe SADC Fisheries project, project report no.40,.

Jul-Larsen E., (1999), Social Capital: Old wine in new bottles? An anthropological assessment, Paper presented at World Bank Seminar on Social Capital, Chr. Michelsen Institute, June, (mimeo).

Jul-Larsen E., Bukali da Graca, Raakjær Nielsen J. & Zwieten P. van, (1997), Research and fisheries management; the uneasy relationship. Review of the Zambia – Zimbabwe SADC fisheries project, Bergen, Chr. Michelsen Institute.

Jul-Larsen E. & B. Kassibo, (2001), Fishing at home and abroad; access to waters in Niger's Central Delta and the effects of work migration, in: *Politics, Property and Production. Natural Resources Management in the Sahel*, Benjaminsen T.A., Lund C. (eds.), Uppsala, Nordiska Afrika Institutet: 208-232.

Jul-Larsen E., (2003), Analysis of effort dynamics in the Zambian inshore fisheries of Lake Kariba, in *Management, co-management or no-management? Major dilemmas in southern African freshwater fisheries*, E. Jul-Larsen, J. Kolding, R. Overå, J. Raakjær Nielsen & P.A.M. van Zwieten, FAO Fisheries Technical Paper 426/2. Rome, FAO.

Jørgensen T., (1990), Long-term changes in age at sexual maturity of Northeast Arctic cod (Gadus morhua L.), *J. Cons. int. Explor. Mer* 46: 235-248.

Kalk M., McLachlan A.J., Howard-Williams C., (1979), *Lake Chilwa: studies of change in a tropical ecosystem*, The Hague, Junk.

Karenge L.P. & Kolding J., (1995a), On the relationship between hydrology and fisheries in Lake Kariba, central Africa, *Fish. Res.* 22: 205-226.

Karenge L.P. & Kolding, J., (1995b), Inshore fish population changes in Lake Kariba, Zimbabwe, in *Impact of Species Changes in African Lakes*, T.J. Pitcher & P.J.B. Hart (eds.), London, Chapman & Hall: 245-275.

Kenmuir D.H.S., (1982), Fish production prospects in Zimbabwe, *Zimbabwe Agric. J.* 79(1):11-17.

Kimpe P.D., (1964), Contribution a l'étude hydrobiologique du Luapula-Moero, Tervuren, Musee Royal de l'Afrique Centrale de Belgique.

Kolding J., (1989), The Fish Resources of Lake Turkana and Their Environment. Cand. Scient. thesis, B, Dept. of Fisheries Biology, University of Bergen.

Kolding J., (1992), A summary of Lake Turkana: an ever-changing mixed environment, *Mitt. Int. Verein. Limnol.* 23: 25-35.

Kolding J., (1993a), Population dynamics and life history styles of Nile tilapia (Oreochromis niloticus) in Fergusons Gulf, Lake Turkana, Kenya, *Env. Biol. Fish.* 37: 25-46.

Kolding J., (1993b), Trophic interrelationships and community structure at two different periods of Lake Turkana, Kenya - a comparison using the ECOPATH II box model, in *Trophic models of aquatic ecosystems*, V. Christensen & D. Pauly (eds.), ICLARM Conference Proceedings No. 26, Manilla, ICLARM: 116-123.

Kolding J., (1994), Plus ça change, plus c'est la même chose. On the ecology and exploitation of fish in fluctuating tropical environments, Dr. Scient. thesis, Bergen, Department of Fisheries and Marine Biology, University of Bergen.

Kolding J., (1995), Changes in species composition and abundance of fish populations in Lake Turkana, Kenya, in *Impact of Species Changes in African Lakes*, T.J. Pitcher & P.J.B. Hart (eds.), London, Chapman & Hall: 335-363.

Kolding J., (1997), Diversity, Disturbance and Dubious Dogma. On some difficult concepts in community ecology, in *Naturvitere Filosoferer*, R. Strand & G. Bristow (eds), Megaloceros in cooperation with Centre for the Studies of Science and the Humanities, University of Bergen: 122-141.

Kolding J., H. Ticheler & B. Chanda, (1996), Assessment of the Bangweulu Swamps fisheries. Final Report prepared for WWF Bangweulu Wetlands Project, SNV/Netherlands Development Organisation, and Department of Fisheries, Zambia.

Kolding J., Musando B. & Songore N. (2003a), Inshore fisheries and fish population changes and in Lake Kariba, in *Management, co-management or no-management? Major dilemmas in southern African freshwater fisheries*, E. Jul-Larsen, J. Kolding, R. Overå, J. Raakjær Nielsen & P.A.M. van Zwieten, FAO Fisheries Technical Paper 426/2. Rome, FAO.

Kolding J., H. Ticheler & B. Chanda, (2003b), The Bangweulu Swamp Fisheries, in *Management, co-management or no-management? Major dilemmas in southern African freshwater fisheries*, E. Jul-Larsen, J. Kolding, R. Overå, J. Raakjær Nielsen & P.A.M. van Zwieten, FAO Fisheries Technical Paper 426/2. Rome, FAO.

Kolding J. & Songore N., (in prep.), Fish diversity, stability and disturbance in Lake Kariba 30 years after impoundment.

Kozlowski J., (1980), Density dependence, the logistic equation, and r- and K-selection: A critique and an alternative approach, *Evolutionary Theory* 5: 89-101.

Krebs C.J., (1985), Ecology: The experimental analysis of distribution and abundance, 3rd edition, New York, Harper & Row Publisher.

Lamberth S.J., Bennett B.A. & Clark B.M., (1995) The vulnerability of fish to capture by commercial beach-seine nets in False bay, South Africa, *S. Afr. J. Mar. Sci.* 15: 25-31.

Lamberth S.J., Bennett B.A., Clark B.M. & Janssens, P.M., (1995), The impact of beach-seine netting on the benthic flora and fauna of False bay, South Africa, *S. Afr. J. Mar. Sci.* 15: 115-122.
Lamberth S.J., Clark B.M. & Bennett B.A., (1995), Seasonality of beach-seine catches in False bay, South Africa, and implications for management, *S. Afr. J. Mar. Sci.* 15:157-167.

Langenhove G.V., Amakali M., Bruine B.D., (1998), Variability in flow regimes in Namibian rivers: natural and human induced causes, in: *Water Resources Variability in Africa during the XXth Century*, Abidjan, Côte d'Ivoire, IAHS, City: 455 - 462.

Laraque A., Mahe G., Orange D. & Marieu B., (2001), Spatiotemporal variations in hydrological regimes within Central Africa during the XXth century, *Journal of Hydrology* 245: 104-117.

Larkin, P.A., (1977), An Epitaph for the Concept Maximum Sustainable Yield, *Trans. Amer. Fish. Soc.* 196: 1-11.

Law R. & Grey D.R., (1988), Maximum Sustainable Yields and the Self-Renewal of Exploited Populations with Age-Dependent Vital Rates, in *Size-Structured Populations*, B. Ebenman & L. Persson (eds.) Berlin, Springer Verlag: 140-154.

Leach M., R. Mearns & I. Scoones, (1999), Environmental entitlements: dynamics and institutions in community-based natural resource management, *World Development* 27: 225-247.

Le Barbe L. & Lebel T., (1997), Rainfall climatology of the HAPEX-Sahel region during the years 1950-1990. *Journal of Hydrology* 188-189: 43-73.

Le Cren E.D. & Lowe-McConnell R.H. (eds.), (1980), *The functioning of freshwater ecosystems*, International Biological Programme 22, Cambridge, Cambridge University Press.

Lehman C.L. & Tilman D., (2000), Biodiversity, stability, and productivity in competitive communities, *American Naturalist* 156: 534-552.

Lévêque, C., (1995), Role and consequences of fish diversity in the functioning of African freshwater ecosystems: a review, *Aquatic Living Resources* 8: 59-78.

Lévêque, C., (1997), *Biodiversity, dynamics and conservation: the freshwater fish of tropical Africa*, Cambridge, Cambridge University Press.

Levinton J.S., (1982), *Marine Ecology*, Englewood Cliffs, N.J., Prentice-Hall Inc.

Lindeman R.L., (1942), The trophic-dynamic aspect of ecology, *Ecology* 23: 399-418.
Lowe-McConnel R.H., (1987), *Ecological studies in tropical fish communities*, Cambridge, Cambridge University Press.

MacArthur R.H. & Wilson E.O., (1967), *The Theory of Island Biogeography*, Monographs in Population Biology 1, New Jersey, Princeton University Press.

Machena C. & Mabaye A.B.E. (1987), Some management aspects and constraints in the Lake Kariba fishery, *NAGA, ICLARM Q.* 10(4): 10-12.

Magnet C., J.E. Reynolds & H. Bru, (2000), A proposal for implementation of the Lake Tanganyika Framework Fisheries Management Plan, GCP/INT/648/NOR-Field Report No. F-14, Rome, FAO.

Malasha I., (2003), The emergence of colonial and post-colonial fisheries regulations: the case of Zambia and Zimbabwe, in *Management, co-management or no-management? Major dilemmas in southern African freshwater fisheries*, E. Jul-Larsen, J. Kolding, R. Overå, J. Raakjær Nielsen & P.A.M. van Zwieten, FAO Fisheries Technical Paper 426/2. Rome, FAO.

Malasha I., (forthcoming), Fisheries co-management: a comparative analysis of the Zambian and Zimbabwean inshore fisheries of Lake Kariba, D.Phil. thesis, Centre for Applied Social Sciences, University of Zimbabwe.

Mandala E.C., (1990) *Work and control in a peasant economy: a history of the Lower Tchiri Valley in Malawi, 1859-1960*, Madison, University of Wisconsin Press.

Margalef R., (1968), *Perspectives in ecological theory*, Chicago, University of Chicago Press.

Margalef R., (1969), Diversity and stability: a practical proposal and a model of interdependence, in *Diversity and stability in ecological systems*, Brookhaven Symposia in Biology No. 22, New York, Brookhaven Nat. Lab.: 25-37

Marshall B.E., (1981), A review of the sardine fishery in Lake Kariba, 1973-1980, Lake Kariba Fisheries Research Institute Project Report No. 40, Harare, Dept. of Nat. Parks and Wildlife Man.

Marshall B.E., (1984), Kariba (Zimbabwe/Zambia) in *Status of African reservoir fisheries*, J.M. Kapetsky & T. Petr (eds.). CIFA Tech. Pap. 10. Rome, FAO: 105-153.

Marshall B.E., (1985), A study of the population dynamics, production and potential yield of the sardine Limnothrissa miodon (Boulenger) in Lake Kariba, PhD thesis, University of Rhodes.

Marshall B.E., Junor F.J.R. & Langerman J.D., (1982), Fisheries and Fish production on the Zimbabwean side of Lake Kariba, *Kariba studies* 10:175-271.

Marshall B.E. & Langerman J.D., (1988), A preliminary reappraisal of the biomass of inshore fish stocks in Lake Kariba, *Fish. Res.* 6:191-199.

Marshall B.E., (1992), The relationship between catch and effort and its effect on predicted yields of sardine fisheries in Lake Kariba. *NAGA, ICLARM Q.* 15(4): 36-39.

May R.M., (1975), Stability in ecosystems: some comments, in *Unifying concepts in ecology*, van Dobben W.H. & Lowe-McConnell R.H. (eds.), Wageningen, Dr W. Junk Publishers: 161-168.

May R.M., (1991), The Chaotic rhythms of life, in: *The New Scientist Guide to Chaos*, N. Hall (ed.), Penguin books: 82-95.

May R.M. (ed.), (1984), *Exploitation of Marine communities. Report on the Dahlem workshop on exploitation of marine communities.* Berlin 1984 April 1-6, Berlin, Springer Verlag.

May R.M., Beddington J.R., Clark C.W., Holt S.J. and Laws R.M., (1979), Management of Multispecies Fisheries, *Science* 205(4403): 267-277.

Maynard-Smith J., (1978), Optimization Theory in Evolution, *Ann. Rev. Ecol. Syst.* 9: 31-56.

Mbewe M. 2000, Impact of Kapenta (Limnothrissa miodon) introduction on the fish community in Lake Itezhi-tezhi, MPhil thesis, Dept. of Fisheries and Marine Biol., University of Bergen.

Mbuga J.S. & A. Getabu, A. Asila, M. Medard, R. Abila, (1998), *Trawling in Lake Victoria, Its history, status and effects*, report no.3, Socio-economics of Lake Victoria, Nairobi, IUCN.

McCracken K.J., (1987), Fishing and the colonial economy: the case of Malawi, in *Journal of African History* 27: 413-429.

Melack J.M., (1976), Primary productivity and fish yields in tropical lakes, *Trans. Am. Fish. Soc.* 105: 575-580.

Medley P.A.H., Gaudian G. & Wells S., (1993), Coral reef fisheries stock assessment, *Reviews in Fish Biology and Fisheries* 3: 242-285.

Mertz D.B. & Wade M.J., (1976), The prudent prey and the prudent predator, *American Naturalist* 110: 489-496.

Misund O.A., Kolding J. & Fréon P., (in press), Fish capture devices and their influence on fisheries management. Chapter 3 in *Handbook on Fish and Fisheries*, vol. II, P.J.B. Hart & J.D. Reynolds (eds.). London, Blackwell Science.

Mosley P., Harrington J. & Toye J., (1991), *Aid and power; the World Bank and policy-based lending*, 2 vols., London and New York, Routledge.

Moyo N.A.G., (1990), The inshore fish yield potential of Lake Kariba, Zimbabwe, *Afr. J. Ecol.* 28: 227-233.

Murphy G.I., (1968), Pattern in Life-History and the Environment, *American Naturalist* 102: 391-403.

Mwakiyongo K.R. & Weyl O.L.F., (2002), Management Recommendations for the Nkacha net fishery of Lake Malombe, in: *Lake Malawi Fisheries Management Symposium*, Weyl, O.L.F., Weyl, M. (eds.), Lilongwe, City.

Nagelkerke L.A.J., Mina M.V., Wudneh T., Sibbing F.A. & Osse J.W.M., (1995), Lake Tana: a unique fish fauna needs protection, *BioScience* 45: 772-775.

Nédélec C. (ed.), (1975), *Catalogue of Small-Scale Fishing Gear*, Oxford, Fishing News Books.

Nédélec C. & Prado P., (1990), *Definition and classification of fishing gear categories*, Fisheries Technical Paper No. 222 Rev.1, Rome, FAO.

Neis B., Schneider D.C., Felt L., Haedrich R.L., Fisher J. & Hutchings J.A., (1999), Fisheries assessment: what can be learned from interviewing resource users? *Can. J. Fish. Aquat. Sci.* 56: 1949-1963.

Nicholson S.E., (1996), A review of climate dynamics and climate variability in Eastern Africa, in *The limnology, climatology and paleoclimatology of teh East African lakes*, Johnson T.C., Odada E. (eds.), Gordon & Breach: 25 - 56.

Nicholson S.E., (1998a), Historical fluctuations of Lake Victoria and other lakes in the Northern Rift Valley of East Africa, in *Environmental change and response in East African lakes*, Lehman J.T. (ed.), Dordrecht, Kluwer Academic Publishers: 7 - 36.

Nicholson S.E., (1998b), Historical fluctuations of rift valley lakes Malawi and Chilwa during historical times: a synthesis of geological, archaeological and historical information, in *Environmental change and response in East African lakes*, Lehman, J.T. (ed.) Dordrecht, Kluwer Academic Publishers: 207 - 231.

Nicholson S.E., (1999), Historical and modern fluctuations of Lakes Tanganyika and Rukwa and their relationship to rainfall variability, *Climatic Change* 41, 53-71.

Nicholson S.E. & Yin X., (1998), *Variations of African lakes during the last two centuries*, IAHS-AISH Publication: 181-187.

Nicholson S.E. & Yin X., (2001), Rainfall conditions in equatorial East Africa during the nineteenth century as inferred from the record of Lake Victoria, *Climatic Change* 48, 387-398.

Noakes D.L.G. & Balon, E.K., (1982), Life Histories of Tilapias: An evolutionary Perspective, in *The Biology and Culture of Tilapias*, R.S.V. Pullin & R.H.L. Lowe-McConnell (eds.), ICLARM Conference Proceedings 7, Manilla, ICLARM: 61-82.

NORAD, (1985), Evaluation Report on the Lake Turkana Fisheries Development Project, Evaluation Report No. 5/85, Oslo, NORAD.

Normann A.K, J. Raakjær Nielsen & S. Sverdrup-Jensen (eds.), (1998), *Fisheries Co-management in Africa*. Proceedings from a rerional workshop on fisheries co-management research, Fisheries Co-management Research Project Research Report No 12, Hirtshals, IFM.

Odum E.P., (1969), The strategy of ecosystem development, *Science*, 164: 262-270.

Oostenbrugge J.A.E. van, Densen W.L.T. van & Machiels M.A.M., (in prep), How coastal communities in the Central Moluccas, Indonesia, must cope with the uncertain outcome of both aquatic and terrestrial resources.

Oostenbrugge J.A.E. van, Bakker W.J., Densen W.L.T. van, Machiels M.A.M. & Zwieten, P.A.M. van, (submitted to Can.J.Fish.Res.), How effective is a multispecies fishery in reducing the variability in the catch from day-to-day?

Orians G.H., (1975), Diversity, stability and maturity in natural ecosystems, in *Unifying concepts in ecology*, van Dobben, W.H. & Lowe-McConnel R.H. (eds.), The Hague ,W. Junk Publishers: 139- 150.

Ostrom E., (1990), *Governing the commons – the evolution of institutions for collective action*, Cambridge, Cambridge University Press.

Overå R., (1998), Partners and competitors: gendered entrepreneurshipin Ghanian canoe fisheries, Dr. Polit. thesis, Bergen, Dept. of Geography, University of Bergen.

Overå, R., (2003), Market systems and investment "bottlenecks", in *Management, co-management or no-management? Major dilemmas in southern African freshwater fisheries*, E. Jul-Larsen, J. Kolding, R. Overå, J. Raakjær Nielsen & P.A.M. van Zwieten, FAO Fisheries Technical Paper 426/2. Rome, FAO.

Paffen P. & E. Lyimo, 1996, Frame survey results for the Tanzanian coast of Lake Tanganyika, March 1995 and comparison with past surveys, GCP/RAF/271/FIN-TD/49, Mpulungu, FAO.

Paloheimo J.E. & Regier H.A., (1982), Ecological Approaches to Stressed Multispecies Fisheries Resources, in M.C. Mercer (ed.), Multispecies Approaches to Fisheries Management, *Can. Spec. Publ. Fish. Aquat. Sci.* 59: 127-132.

Panayotou, T., (1982). *Management concepts for small-scale fisheries: Economic and social aspects*. FAO Fish. Tech. Pap. 228, Rome, FAO.

Pauly, D., (1997) *Theory and management of tropical multispecies stocks: A review, with emphasis on the Southeast Asian demersal fisheries*. ICLARM Studies and Reviews 1., Manila, ICLARM.

Pauly, D., (1994), On Malthusian overfishing, in *On the sex of fish and the gender of scientists: essays in fisheries science*, D. Pauly, London, Chapman & Hall: 112-117.

Pauly, D., (1979), Small-scale fisheries in the tropics: marginality, marginalization and some implication for fisheries management, in *Global trends: Fisheries Management*, E.K. Pikitch, D.D. Huppert & M.P. Sissenwine (eds.). American Fisheries Society Symposium 20, Bethesda, Maryland: 40-49.

Pearce M., (1995), The removal and return of fishers in the Sinazongwe islands, in Proceedings of the inshore working group workshop on Lake Kariba, Kariba, Zambia-Zimbabwe SADC fisheries project, project report no.45: 50-56.

Pet J.S., (1995), On the management of a tropical reservoir. PhD thesis, Fish Culture and Fisheries, Wageningen, Wageningen, Agricultural University.

Pet J.S., Machiels M.A.M. & Densen, W.L.T. van, (1996), A size-structured simulation model for evaluating managment strategies in gillnet fisheries exploiting spatially differentiated populations, *Ecological Modelling* 88: 195-214.

Peters P. E., (1994), *Dividing the commons; politics, policy and culture in Botswana*, Charlottesville, University Press of Virginia.

Peters P. E., (2000), Grounding governance: Power and meaning in natural resource management, Keynote address to the International Symposium on Contested Resources:

Challenges to Governance of Natural Resource in Southern Africa, University of the Western Cape, October 2000.

Peters R.H., (1991), *A critique for ecology*, Cambridge, Cambridge University Press.

Petraitis P.S., Latham R.E. & Niesenbaum R.A., (1989), The maintenance of species diversity by disturbance, *Quart. Rev. Biol.* 64(4): 393-418.

Pianka E.R., (1970), On r and K selection, *American Naturalist* 104, 592-597.

Pimm S.L., (1984), The complexity and stability of ecosystems, *Nature* 307: 321-326.

Pitt T.K., (1975), Changes in the Abundance and Certain Biological Characters of Grand Bank American Plaice, Hippoglossoides platessoides. *J.Fish. Res. Board Can.* 32: 1383-1398.

Platteau J.-P. & A. Abraham, (1987), An inquiry into quasi-credit contracts: The role of reciprocal credit and interlinked deals in small-scale fishing communities, *Journal of Development Studies* 23 (4): 461-90.

Platteau J.-P., (1989a), The Dynamics of fisheries development in developing countries: A general overview, *Development and Change*, 20 (4): 565-597.

Platteau J.-P., (1989b), Penetration of capitalism and persistence of small-scale organisational forms in third world fisheries. *Development and Change*, 20 (4): 621-651.

Plisnier P.-D., Chitamwebwa D., Mwape L., Tshibangu K., Langenberg V. & Coenen, E., (1999), Limnological annual cycle inferred from physical-chemical fluctuations at three stations of Lake Tanganyika, Hydrobiologia 407: 45-58.

Pope J.G., (1977), Estimation of fishing mortality, its precision and implications for the management of fisheries, in *Fisheries Mathematics*, J.H. Steele (ed.), London, Academic press: 63-76.

Pope J.G., Shepherd J.G. & Webb J., (1994), Successful surf-riding on size spectra: the secret of survival in the sea, *Phil. Trans. R. Soc. Lond.* B 343, 41-49.

Power G. & Gregoire J., (1978), Predation by Freshwater Seals on the Fish Community of Lower Seal Lake, Quebec, *J. Fish. Res. Board Can.* 35: 844-850.

Ramberg L., Björk-Ramberg S., Kautsky N. & Machena C., (1987), Development and biological status of Lake Kariba - A man-made tropical lake, *Ambio* 16: 314-321.

Rapport D.J., Regier H.A. & Hutchinson T.C., (1985), Ecosystem behaviour under stress, *American Naturalist* 125: 617-640.

Regier H.A., (1973), Sequence of exploitation of stocks in multispecies fisheries in the Laurentian great lakes, *Journal of the Fisheries Research Board of Canada* 30: 1992 - 1999.

Regier H.A., (1977), Fish communities and aquatic ecosystems, in *Fish Population Dynamics*, J.A. Gulland (ed.), London, John Wiley & Sons: 134-155.

Regier H.A. & Loftus K.H., (1972), Effects of fisheries exploitation on salmonid communities in oligotrophic lakes, *Journal of the Fisheries Research Board of Canada* 29: 959-968.

Regier H.A. & Henderson H.F., (1972), Towards a broad ecological model of fish communities and fisheries, *Transactions of the American Fisheries Society* 1: 56-72.

Ricker W.E., (1954), Stock and recruitment, *J. Fish. Res. Bd. Can.* 11: 559-623.

Roff D.A., (1984), The evolution of life history parameters in teleosts, *Can J. Fish. Aquat. Sci.* 41: 989-1000.

Rothchild B.J., (1977), Fishing effort, in *Fish Population Dynamics*, J.A. Gulland (ed.), London, John Wiley & Sons: 96-115.

Rothschild B.J. (1986), *Dynamics of Marine Fish Populations*, Cambridge, Massachusetts and London, Harvard University Press.

Sainsbury K.J., (1982), The ecological basis of tropical fisheries management, in *Theory and management of tropical fisheries*, Pauly, D. & G.I. Murphy (eds.), ICLARM Conference proceedings No. 9, Manila, ICLARM: 167-188.

Salz P., (1986), *Policy instruments for development of fisheries*, Publication No. 574, The Hague, Agricultural Economics Research Institute.

Sanford S., (1983), *Management of pastoral development in the Thirld World*, Chichester, Wiley.

Sarvala J., Salonen K., Jaervinen M., Aro E., Huttula T., Kotilainen P., Kurki H., Langenberg V., Mannini P., Peltonen A., Plisnier P.D., Vuorinen I., Moelsae H. & Lindqvist, O.V., (1999), Trophic structure of Lake Tanganyika: carbon flows in the pelagic food web, *Hydrobiologia* 407: 149-173.

Schaefer M.B., (1954), Some aspects of the dynamics of populations important to the management of the commercial marine fisheries, *Bulletin of the Inter-American Tropical Tuna Commission* 1: 27-56.

Schaffer W.M., (1974), Optimal Reproductive Effort in Fluctuating Environments, *American Naturalist* 108: 783-790.

Scholz U., G. Mudenda & H. Möller, (1997), Some aspects of the development of the small-scale fishery on the Zambian side of Lake Kariba 1961-1990 and implications for fisheries management, in *African inland fisheries, aquaculture and the environment,* K. Remane (ed.), Oxford, Fishing News Books: 255-271.

Scoones I. (ed.), (1995), *Living with uncertainty: new directions in pastoral development in Africa*, London, Intermediate Technology Publications.

Scoones I. & B. Cousins, (1994), Struggle for control over wetland resources in Zimbabwe, *Society and Natural Resources*, 7: 579-594.

Scott J.C., (1976), *The moral economy of the peasants: rebellion and subsistence in Southeast Asia*, London, Yale University Press.

Scudder T., (1965), The Kariba case: man-made lakes and resource development in Africa. *Bulletin of the atomic scientists*, December 6-11.

Shanin T. (ed.), (1971), *Peasants and peasant societies*, Harmondsworth, Penguin books.

Sheldon R.W., Prakash A. & Jr. W.H.S., (1972), The size distribution of particles in the ocean, *Limnology and Oceanography* 17: 327-340.

Silliman R.P. & Gutsell J.S., (1958), Experimental exploitation of fish populations. *U.S. Fish Wildl. Serv., Fish. Bull.* 133: 214-252.

Sissenwine M.P., (1978), Is MSY an Adequate Foundation for Optimum Yield? *Fisheries* 3(6): 22-42.

Sissenwine M.P., (1984), Why do fish populations vary? in *Exploitation of marine communities*, R.M. May (ed.) Dahlem workshop 1984, Berlin, Springer-Verlag: 59-94.

Skjønsberg E., (1992), Men, money and fisheries planning: the case of the Northern Province of Zambia. in *Fishing for development*, Tvedten I. & B. Hersoug (eds.), Uppsala, Nordic Africa Institute: 155-172.

Slobodkin L.B., (1964), The strategy of evolution, *Am. Sci.* 52: 342-357.

Slobodkin L.B., (1968), How To Be a Predator, *Am. Zool.* 8: 43-51.

Slobodkin L.B., (1974), Prudent Predation does not Require Group Selection, *American Naturalist* 108: 665-678.

Slobodkin L.B., Smith F.E. & Hairston N.G., (1967), Regulation in terrestrial ecosystems and the implied balance of nature, *American Naturalist* 101: 109-124.

Slobodkin L.B., (1961), *Growth and regulation of animal populations*, New York, Holt, Rinehart & Winston.

Songore N., (2000), Compilation and Analysis of Catch-Effort Data on the Lake Kariba Artisanal Fishery, Zimbabwe, Management, Co-management, or No Management Project, Internal report (mimeo).

Southwood T.R.E., May R.M., Hassell M.P. & Conway G.R., (1974), Ecological Strategies and Population Parameters, *American Naturalist* 108: 791-804.

Spigel R.H., Coulter G.W., (1996), Comparison of hydrology and physical limnology of the East African great lakes: Tanganyika, Malawi, Victoria, Kivu and Turkana (with reference to some North American Great Lakes), in: *The Limnology, Climatology and Paleoclimatology of the East African Lakes*, Johnson, T.C., Odada, E.O. (eds.), Amsterdam, Gordon and Breach Publishers: 103 -140.

Stearns S.C., (1976), Life-History Tactics: A Review of the Ideas, *Q. Rev. Biol.* 51: 3-47.

Stearns S.C., (1977), The Evolution of Life History Traits: A Critique of the Theory and a Review of the Data, *Ann. Rev. Ecol. Syst.* 8: 145-171.

Stearns S.C. & Crandall R.E., (1984), Plasticity for Age and Size at Sexual Maturity: A Life-history Response to Unavoidable Stress, in *Fish Reproduction – Strategies and Tactics*, G.W. Potts & R.J. Wootton (eds.), London, Academic Press: 13-33.

Sutherland W.J., (1990), Evolution and fisheries, *Nature* 344: 814-815.

Sverdrup-Jensen, S. & J. Raakjær Nielsen, (1998), Co-management in small-scale fisheries. A synthesis of Southern and West African experiences. In Normann *et al.* 1998.

Swift J.J., (1988), Major issues in pastoral development with special emphasis on selected African countries, Rome, FAO.

Tanner J.T., (1966), Effects of population density on growth rates of animal populations, *Ecology* 47: 733-745.

Talling J.F., Lemoalle J., (1998), *Ecological dynamics of tropical inland waters*, Cambridge, Cambridge University Press.

Thomson D., (1980), Conflict within the fishing industry, *ICLARM Newsletter*, 3(3): 3-4.

Ticheler H., Kolding J. & Chanda, B., 1998, Participation of local fishermen in scientific fisheries data collection, a case study from the Bangweulu Swamps, Zambia, *Fish. Man. and Ecology*, 5: 81-92.

Tiffen M., M. Mortimore & F. Ghichuki, (1994), *More people, less erosion: environmental recovery in Kenya*, Chichester, Wiley.

Tilman D., Lehman C.L. & Thomson K.T., (1997), Plant diversity and ecosystem productivity: Theoretical considerations in *Proceedings of the national academy of sciences of the United States of America* 94: 1857-1861.

Tilman D., Lehman C.L., Bristow C.E., (1998), Diversity-stability relationships: Statistical inevitability or ecological consequence? *American Naturalist* 151: 277-282.

Tonn W.M. & Magnuson J.J. (1982), Patterns in the species composition and richness of fish assemblages in Northern Wisconsin lakes, *Ecology* 63(4): 1149-1166.

Tweddle D.E., (1995), Proceedings of the Fisheries Research Symposium, November 1994. Fisheries Bulletin 33, Lilongwe, Fisheries Department, Malawian Ministry of Natural Resources.
Underwood A.J., (1989), The analysis of stress in natural populations, *Biological Journal of the Linnean Society*, 37: 51 – 78.

Verheust L. & Johnson G., (1998), The watershed database for sub-equatorial Africa, structure and user interface, ALCOM working paper No. 16, Harare, ALCOM/FAO.

Verhulst P.F, (1838), Notice sur la loi que la population suit dans son acroissement. *Correspondance Math. Phys.* 10: 113-121.

Vetter E.F., (1988), Estimation of natural mortality in fish stocks: a review, *Fish. Bull.* 86(1): 25-43.

Ware D.M., (1975), Relation between egg size, growth, and natural mortality of larval fish, *J. Fish. Res. Bd. Can.* 32: 2503-2512.

Welcomme R.L., (1999), A review of a model for qualitative evaluation of exploitation levels in multispecies fisheries, *Fisheries Management and Ecology* 6: 1-19.

Weyl, O.L.F., (2001). Trends in Malawi Fisheries, (mimeo).

Wilbur H.M., Tinkle D.W. & Collins J.P., (1974), Environmental Uncertainty, Trophic Level and Resource Availability in Life History Evolution, *American Naturalist* 108: 805-817.

Wolf E., (1966), *Peasants*, New Jersey, Prentice-Hall.

Yoccoz N.G., Nichols J.D. & Boulinier T., (2001), Monitoring of biological diversity in space and time, *Trends in Ecology & Evolution* 16, 446-453.

Zwieten P.A.M. van, Aarnink B.H.M. & Kapasa C.K., (1995), How diverse a fishery can be! Structure of the Mweru-Luapula fishery based on an analysis of the Fram Survey 1992 and a characterisation of the present management strategies. DoF/ML/1996 25, Nchelenge, Zambian Department of Fisheries.

Zwieten P.A.M. van, Roest F.C., Machiels M.A.M., Densen W.L.T. van, (2002c), Effects of interannual variability, seasonality and persistence on the perception of long-term trends in catch rates of the industrial pelagic purse-seine fisheries of Northern Lake Tanganyika (Burundi), *Fisheries Research* 54: 329 -348.

Zwieten P.A.M. van & Njaya F., (2003), Environmental variability, effort development and the regenerative capacity of the fish stocks in Lake Chilwa, Malawi, in *Management, co-management or no-management? Major dilemmas in southern African freshwater fisheries*, E. Jul-Larsen, J. Kolding, R. Overå, J. Raakjær Nielsen & P.A.M. van Zwieten, FAO Fisheries Technical Paper 426/2. Rome, FAO.

Zwieten P.A.M. van, Njaya F. & Weyl O.L.F., (2003a), Effort development and the decline of the fisheries of Lake Malombe: does environmental variability matter? in *Management, co-management or no-management? Major dilemmas in southern African freshwater fisheries*, E. Jul-Larsen, J. Kolding, R. Overå, J. Raakjær Nielsen & P.A.M. van Zwieten, FAO Fisheries Technical Paper 426/2. Rome, FAO.

Zwieten P.A.M. van, Goudswaard P.C. & Kapasa C.K., (2003b), Mweru-Luapula is an open exit fishery where a highly dynamic population of fishers makes use of a resilient resource base, in *Management, co-management or no-management? Major dilemmas in southern African freshwater fisheries*, E. Jul-Larsen, J. Kolding, R. Overå, J. Raakjær Nielsen & P.A.M. van Zwieten, FAO Fisheries Technical Paper 426/2. Rome, FAO.

Zwieten P.A.M. van, Densen, W.L.T. van, Dang Van Thi, (2002d). Improving the usage of the fisheries statistics of Vietnam for production planning, fisheries management and nature conservation, Marine Policy 26(1): 13-34.

Appendix 1

THEORETICAL ELABORATION ON ECOLOGICAL CONCEPTS AND THE DEVELOPMENT OF FISHING PATTERNS IN MULTISPECIES FISHERIES

Section 1. (Surplus) production, productivity, trophic level and density dependence

Biological production (*P*) is the total amount of tissue generated in a population in a particular space during a given period of time. It is of central interest in the exploitation of renewable resources, since the yield (*C*) is a fraction *F* (Fishing mortality) of the mean biomass and is a fraction (*x*) of biological production *P*.

$$C = F \cdot \overline{B} = x \cdot P \tag{1}$$

Production thus includes both living organisms and organisms that died within the time period. Gains in biomass are a result of individual growth, new offspring and immigration, whereas death and emigration cause losses in biomass.

'Surplus' production, or the net production after natural density-dependent mortality has been subtracted, is essential for any population in order to expand and/or withstand predation without declining. The *ecotrophic efficiency* (*x* in eqn. 1) is defined as the fraction of total production that is consumed by higher trophic levels. The concept of *trophic level* (Lindeman, 1942) means grouping of taxa or populations into discrete levels according to their place in a food chain, e.g. primary producers (plants and algae), herbivores, first-order carnivores, second-order carnivores, etc. This system is used to simplify the description of an ecosystem, but also to describe the interactions and efficiencies of energy transfers between trophic levels. Most models used in fisheries are single-species models with only two 'trophic levels', the stock as prey and the fisherman as the only predator.

The concept of Maximum Sustainable Yield (MSY), that is the maximum surplus production, was derived from sigmoid curve theory (Graham, 1935). The theory describes the change in biomass and production within a population. It presupposes that the regeneration of biomass, or net rate of increase, is a density-dependent function of biomass (*dB/dt = g(B)*) which is dome-shaped with its highest point (= MSY) at some intermediate level between 0 and a maximum density. The simplest mathematical model for self-regulating growth in populations is the well-known logistic equation (Verhulst, 1838). Although widely criticized for oversimplification (e.g. Kozlowski, 1980), it has contributed much to ecological thinking and forming of ideas. Its greatest applicability is perhaps in the illustration of the theory of density-dependent growth, which is best shown in its differential dome-shaped form

$$g(B) = \frac{dB}{dt} = r_m B \cdot \left(1 - \frac{B}{K}\right) \tag{2}$$

In this equation, *K* is the theoretical maximum biomass (*B*) that can be attained in an environment (= B_{max} or B_∞) also known as carrying capacity, which is mainly determined by available food and space. The parameter r_m is the intrinsic rate of natural increase in biomass. It is defined as the maximum instantaneous rate of births per individual (*b*), minus the minimum instantaneous rate of natural deaths per individual (*d*) under specific environmental conditions or $r_m = b - d$. (Fenchel, 1977). The logistic growth described by (2) starts with an exponential growth phase of a population that decreases as the relative saturation of the environment increases, until the asymptote K is reached (i.e. where $K = B$) and growth ceases ($g(B)=0$). This means that the per capita, or actual rate of increase (r) in the population is a linear function of density (B) with $r = r_m$ at the intercept, which is seen when rewriting (2):

$$r = \frac{dB}{B \cdot dt} = r_m - cB \qquad (3)$$

If we then assume that the potential birth rate (fecundity) is constant (Ricker, 1954; Beverton and Holt, 1957) then the death rate must be a linear function of the density too, since $d = b - r_m$. The assumption of density-independent birth rate may not be true for the full range of r. However, it is generally accepted at lower population densities, while it seems valid at least up to the inflection point (K/2 in (2)) where intra-specific competition for resources starts affecting productivity (Begon *et al.*, 1990, p. 202).

Production is thus a density-dependent *quantity* expressed in kilograms or tonnes, often scaled by area or volume. Productivity is the rate or *speed* at which production is generated and is a function of both the individual biological regenerative characteristics of a particular species (the per capita rate of increase), and the density (B) of the stock. Productivity is the instantaneous rate of biomass production dB/dt. In the presence of fishing, the instantaneous rate of change in biomass is equal to the productivity minus the accrual rate of the yield, or, combining the basic fisheries equations C(atch)=F(ishing mortality)·B(iomass) and F=q·f =catchability·effort and (2):

$$\frac{dB}{dt} = g(B) - F \cdot B = g(B) - q \cdot f \cdot B \qquad (4)$$

In words, what this equation says is that, over a fixed period of time, the change in biomass is the surplus production minus the yield:

new biomass = old biomass + surplus production – catch

When biomass does not change ($dB/dt = 0$), then surplus is equal to output and a stock is said to be at an equilibrium. Reasoning along these lines has led to a set of important fishery models called surplus production models, of which the Schaefer (1954) model (equivalent to (2)) is an example. Another important inference is that at any constant population size, the average total death rate (Z) is equal to the intrinsic rate of increase (r_m), which implies that the total mortality rate of a population is related to this important life history characteristic of the species (Kolding, 1994).

Productivity is also used more loosely in this report for example when "changes in lake productivity" are related to "changes in fish yield". What is meant by these expressions is that, through environmentally driven processes such as changes in water inflow in lakes and resulting water levels, the productivity at different trophic levels changes, resulting in changes in fish production and hence fish yields. Therefore, in a changing environment, the idea of a constant carrying capacity (which is the underlying assumption for (2)) is neither plausible nor necessary for conceiving density dependence or equilibrium situations. Steady state means that the actual rate of increase $r = 0$. In a growing population $r > 0$ and in a declining population $r < 0$. Consequently, in nature the value of r for all non-extinct populations is fluctuating around a mean value of 0. It also implies that there will always be a set of environmental conditions at which r is positive. A theoretical set even exists where r attains a maximum value (r_m) at steady state conditions, leading back to equation (2). Since r depends on the age structure in the population, it is clear that any specific value of r is only valid for a particular environment and mortality regime. The frequency and amplitude of oscillations in r must then be mainly related to the variability of the environment (the extrinsic factors). In fish, the change in r is dependent both on abiotic factors, such as temperature and oxygen, and biotic factors such as food and predation.

All factors limiting population growth are considered to be density dependent (Andrewartha and Birch, 1954). In order of increasing importance these factors are:

- Shortage of resources like food and shelter,
- Inaccessibility of these resources in relation to the animal's capacity of dispersal, and
- Shortages of time duration in which the rate of increase (r) is positive.

Climate or predators influence the fluctuations in r, the last, most important factor. Depending on the environment, one or the other has overwhelming importance. The occurrence of intermittent short spells of optimal situations might well be illustrated by the strong year-class variation we observe in most fisheries. The huge variations are considered one of the biggest obstacles in fisheries modelling where traditionally, an attempt is made to relate recruitment with stock size. Yet stock-recruitment theory, as emphasized by Rothschild (1986), is simply a theory that attempts to account for the mortality of young fish between spawning and recruitment time. Thus we may generalize to two situations: one in which population size is largely determined by climate, and another where other animals largely determine it. In both cases mortality is still the most important common denominator.

Section 2. Diversity, stability, resilience and regenerative capacity

Fishing, predation and environmental changes all causes stress, and the capacity to recover, persist, endure or 'bounce back' to a previous state, is theoretically associated with the two concepts: *stability* and *resilience*. According to one of the better known definitions (Holling, 1973), stability is "The ability of a system to return to an equilibrium state after a temporary disturbance, the more rapidly it returns and the less it fluctuates, the more stable it would be". Again according to Holling, (op, cit.) resilience is: "A measure of persistence of systems and the ability to absorb change and disturbance and still maintain the same relationships and composition between populations or state variables [irrespective of relative abundance]". In Holling's view, instability in the sense of large fluctuations, produces a highly resilient system capable of repeating itself and persisting over time until a disturbance restarts the sequence. Thus systems can be very resilient and still fluctuate wildly. Holling (op. cit.) states that these two distinct properties alone define the behaviour of ecological systems. However, any measure of stability requires a time-perspective that must be seen in relation to the lifespan of a species. The ambiguities in the literature between stability and resilience might thus, like the equilibrium and non-equilibrium models discussed later, simply be a matter of scale (Kolding, 1997). Therefore the mortality rate and pattern represents the speed of regeneration against which stability and resilience should be measured. As shown in Figure 5.7 'stable' species have on average a low intrinsic growth rate (r) and total mortality (Z) with a correspondingly longer lifespan, whereas 'resilient' species have a high r and Z and shorter lifespan. What determines resilience and stability depends on the combination of stress (continuous selective or discrete non-selective mortality), and the trade-off between the advantages of being big, or developing specialized behaviour, and the probability of dying with time.

The coexistence of several species in an ecosystem, so-called biodiversity, and particularly the natural regulation and maintenance of biodiversity is theoretically a challenge (Kolding, 1997). The basic unit in biodiversity is the individual species and normally the focus is on the number of species and the relative abundance and distribution of individuals within an ecosystem (*alpha* diversity). The more species, the more diverse is the ecosystem, and the more we tend to value it. Consequently, fisheries are facing a dilemma against the drive to conserve biodiversity. For instance, FAO (1992, p. 5) wrote: "Continued high fishing intensity will contribute to a loss of biological diversity,(...) and this may lead to more unstable, and possibly lower, catches in the long term". From the background assumption of "The balance of nature" (Egerton, 1973), system complexity, diversity and environmental stability have traditionally been positively related to each other (Margalef, 1968, 1969; Odum, 1969). This 'diversity-stability' hypothesis has often led to the suggestion that highly diverse communities are particularly vulnerable to exploitation (May, 1975; Sainsbury, 1982).

In contrast, theoretical advancement in understanding species diversity generation, notably the 'intermediate disturbance hypothesis' (Connell, 1978), and the 'dynamic equilibrium hypothesis' (Huston, 1979), build on non-equilibrium dynamic community assumptions. Both infer that frequent but irregular disruptions are a major agent in maintaining high-diversity communities. In general, the various hypotheses for the regulation of diversity can be grouped into so-called equilibrium and non-equilibrium models (Tonn and Magnuson, 1982; Petraitis *et al.*, 1989; Begon *et al.*, 1990). Selective, density-dependent, predator-induced mortality belongs to the first category, whereas catastrophic, non-predictive, density-independent, environment-induced mass mortality belongs to the latter. However, common to all these hypotheses is that population reduction in the form of either selective (predation) or non-selective mortality (environmental disturbances) is the main mechanism for the regulation of diversity. The logic is that individual population densities must be (and are, see section 3) kept lower than the carrying capacity to prevent the effects of strong mutual interactions, the so-called *competitive exclusion principle* (Hardin, 1960). Both the selective mortality based hypotheses (equilibrium models), and the hypotheses based on non-selective population reductions (non-equilibrium models) predict the highest diversity to be at an intermediate level of predation, stress or disturbances, i.e. the various populations never gain enough dominance to competitively exclude others. Thus the difference between the two groups of models can simply be reduced to a situation where the population reductions are either continuous or discrete (Kolding, 1997). In other words, the creation and maintenance of biodiversity can be considered to be regulated by the mortality pattern in the ecosystem.

In summary, we can generalize these ecological concepts and processes into two broad categories where the environment determines the prevailing mortality pattern:

- The unstable environment, characterized by discrete, density-independent, non-predictive, non-selective mortality induced by physical changes
- The stable environment, characterized by continuous, density-dependent, predictive, and size-selective mortality induced by the biotic community.

The two broad categories represent extremes on a gradient, and (in)stability must be seen as a time function in relation to the mean generation time of populations. Thus, even for the 'unstable' environment, there are two life-history strategies: either follow the fluctuations (boom-and-bust ephemeral species), or endure the disturbances (long-lived resistant species).

For the latter, the environment may even no longer be unstable, only periodic (Kolding, 1994).

Section 3. The regulation of populations and mortality as a key parameter

Fishing activity is but one of many stress factors to a population. If we can understand the adaptations and life history traits of a population to resist natural mortality factors, we can also evaluate the effect of fishing on these stocks. The diversity and abundance of natural populations is maintained and regulated through a series of interacting factors and associated fundamental concepts in population and community ecology such as: *density dependence, compensatory mechanisms, stability* and *resilience*. There is a distinction between internal processes that are regulated by the abundance of the population itself, such as density dependence (see section 1), and external processes that are controlled by the surrounding environment and community of other species. Without compensatory properties, a population in a density-controlled multi-species system exposed to long-term increased mortality from predation or fishing, would ultimately perish. Most theories on population and community ecology and life histories can be reduced to show that the processes they aim to explain are closely associated with the pattern and rate of mortality (Kolding, 1994, 1997). In essence, it is the transience of life, not life itself, that is the driving force of evolution, simply because dying is more certain than giving birth.

Probably few, if any, natural animal populations utilize or occupy their environment to carrying capacity (Andrewartha and Birch, 1954; Slobodkin *et al.*, 1967; Stearns, 1977). Species will mostly either compete for the resources, or be predators. The influence that species have on each other is difficult to measure. On the other hand, if a competitor or predator is removed from the system and we then observe an expansion of other species, then competition or predation is demonstrated. Such multispecies interactions have been observed in the North Sea (Andersen and Ursin, 1977), in the Antarctic (May *et al.*, 1979), in the Gulf of Thailand (Pauly, 1979), in West Africa (Gulland and Garcia, 1984) and in many freshwater fisheries (Paloheimo and Regier, 1982; Carpenter *et al.*, 1985), where heavy fishing pressure on larger slower-growing species leads to an expansion of smaller faster-growing organisms.

Comparing these observations with the tenets that:

- predation is believed to be the most important factor for natural mortality in fish (Sissenwine, 1984; Vetter, 1988; ICES, 1988);
- adaptations tend to maximize fitness through optimal utilization of resources (Slobodkin, 1974; Stearns, 1976; Maynard-Smith, 1978);
- predators and prey are co-evolved (Slobodkin, 1974; Krebs, 1985);
- there is an uni-modal response of prey productivity to predator densities (sigmoid curve theory, section.1),

it is reasonable to presume that predation in the long term would 'maintain' prey populations close to their highest average surplus production rate (Slobodkin, 1961, 1968; Mertz and Wade, 1976; Pauly, 1979; Caddy and Csirke, 1983; Carpenter *et al.*, 1985). The argument follows simply from the sigmoid curve where the highest sustainable surplus production of the prey population (dB/dt = max = MSY) is also the *'carrying capacit"* (K) of the predator population. The predators can in theory grow to reach K (= MSY_{prey}), but if they overshoot, they will reduce the net prey production and consequently decline themselves from starvation. Stable 'equilibria' in such cybernetic density-controlled predator-prey relations are theoretically only possible up to the inflexion point of the sigmoid growth curve of the prey where dB/dt is maximized.

Any additional mortality at this stage (as in time-lagged predator-prey oscillations), however, requires a change in the life history strategy if the prey is not to perish (Slobodkin, 1974). In other words, when a population adapted to a relatively stable environment is submitted to more long-term changes in the external mortality forces, it must somehow respond by increasing the intrinsic growth rate (r) (Roff, 1984). This requires stress response or compensatory mechanisms (intrinsic changes) which again are related to phenotypic plasticity, a trait that is particularly prominent in fish (Stearns, 1977; Stearns and Crandall, 1984).

r- K selection and size-specific mortality

In evolutionary terms, changes in the survival rate are less efficient in improving r than increasing the turnover rate by decreasing the generation time. Empirical studies have shown that there is a strong inverse correlation between age at maturity and mortality, which can be considered as a trade-off between the advantages of being big and the probability of dying with time (Adams, 1980; Gunderson, 1980; Hoenig, 1983; Roff, 1984; Gunderson and Dygert, 1988). Traditionally, the explanation of this phenomenon was based on the well-known theory of r- and K selection (MacArthur and Wilson, 1967; Pianka, 1970, 1972; Southwood *et al.*, 1974; Boyce, 1984). This theory was associated with the environmental stability, or rather the degree of 'saturation' (density) a population can reach in relation to fluctuating resources. However, considering the indefinable relationship between the carrying capacity (K) and life history traits (Stearns, 1977; Kozlowski, 1980), the original interpretation of the r-K selection is in many ways an inadequate explanation. Other authors (Murphy, 1968; Schaffer, 1974; Wilbur *et al.*, 1974; Stearns, 1977; Horn, 1978) have therefore suggested that the different life-history styles should be considered a function of relative size-specific mortalities. In essence: abiotic mortality, caused by the physical instability of the environment, is generally considered to influence the whole age structure of the population of a species. Thus a low-somatic and high-reproductive allocation of energy indicates that continued existence of the individual beyond the first reproduction is not profitable due to the risk of dying from physical disturbances. On the other hand, biotic mortality (mainly predation) is considered the factor most affecting the small/young individuals in a population (Cushing, 1974; Ware, 1975; Bailey and Houde, 1989; Caddy, 1991). Hence, if mortality is reduced with increasing size, it is advantageous to initially invest more in growth relative to reproduction. Empirically, this is corroborated by 'Copes rule' which states that in the evolution of relatively stable ecosystems, there will be a tendency towards the development of larger sizes within the food-chains (Pianka, 1970; Dickie, 1972; Begon *et al.*, 1990). In conclusion, the balance between reproduction and growth, in an optimal life history, seems determined by the relation between adult and juvenile survival (Charnov and Schaffer, 1973; Horn, 1978).

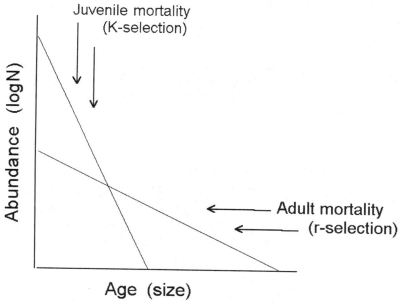

Figure 1 *Selection pressures. A theoretical illustration of the 'r-K selection principle' as a compensatory mechanism for size-specific mortality pressures only. The slopes of the lines are equal to total mortality (Z) assuming simple exponential decay. Under steady-state conditions, the intrinsic rate of increase (r_m) equals Z equals P/B ratio. Arrows indicate the two selection directions and resulting 'strategy'. A high Z-value (steep slope) thus represents a high r-value and a low Z equals a low r.*

Reproduced from Kolding (1993).

From this interpretation of the *r*-K selection principle, then theoretically, even for 'K-selected' species, a compensatory strategy against increased mortality on the adult stages would be to increase the turnover rate (P/B=Z) by reducing the generation time (Figure 5.7). This has been corroborated by empirical studies for American plaice (Pitt, 1975), Lake trout (Power and Gregoire, 1978), Northeast Arctic Cod (Jørgensen, 1990) and Nile tilapia (Kolding, 1993a) that all showed a decreased age of maturity at increased adult mortality. Thus density (section 1), individual size, generation time, and changes in these attributes over time, are all seemingly close functions of death rates in a population. The product of density and size gives the biomass, and the integration of biomass over time gives the production. A further condensation of biomass and production into the P/B ratio then directly reflects the mortality rate and vice versa.

P/B ratio and turnover rate

Production processes are usually associated with the rates at which biological tissue moves between trophic levels, and are thus dynamic quantities, which can rarely be measured directly. The production/biomass ratio (P/B), however, is one way of envisioning the timescale, by indicating the turnover rate and thus the speed of the biomass regeneration. Dickie (1972) emphasized the central importance of this concept for understanding ecological and production efficiencies in relation to fishing pressure. The P/B ratio tends to decrease from one trophic level to another with distance from the primary production level, and also tends to have a general non-linear relationship with the sizes of organisms involved. This means that changes in size-composition of a population from human exploitation or predation will be reflected in the P/B ratio by a relative change in the generation time. The P/B ratio is thus an extremely useful parameter to comparatively characterize different systems, species or trophic levels within a system (Le Cren and Lowe-McConnell, 1980). Allen (1971) examined the P/B ratio for a number of mathematical models expressing mortality and growth. He found that for any growth model (except simple exponential), and with a simple exponential death rate, the P/B ratio is equal to the total instantaneous mortality rate (Z). Thus, the gross production per unit time, P = BZ, is entirely a function of the mean biomass and average mortality rate.

Section 4. The optimal exploitation rate and exploitation pattern

In a fish community with several trophic levels, the amount of production, the speed at which it is generated, and the way it is dispersed through the food-web, determine the production that can be harvested. For fisheries management, the most important implication of density-dependent limitations to growth is that a fishery must substitute one form of mortality for another if the abundance is to remain stable, as the yield is simply the fished fraction of the total deaths. Consequently, in the traditional single-species production models (section 1), a reduction of stock size (from the theoretical $B\infty$ at the carrying capacity K) is the prerequisite for increasing the 'surplus', and calculations are aimed at estimating the point of highest net regeneration rate (MSY). However in a multispecies situation, if natural predation is already harvesting the resource close to this rate (section 3), then a fishery is an additional uncompensated source of mortality and the population is driven to collapse. Fortunately, predation mortality is in practice simultaneously alleviated, as few fisheries are focused on one single species and the predators are also being harvested. In fact, the top levels in an ecosystem are often the first to be exploited intensively (Regier, 1977; Beddington, 1984). Management questions are then:

- how much of the production can be harvested (the exploitation *rate*)?
- what is a rational harvesting strategy or exploitation *pattern* on the community, that is, what rates should be applied to each stock?

These proportions (the optimal exploitation rate and pattern) are complicated in a multispecies situation (Dickie, 1972; May *et al.*, 1979; Beddington and May, 1982; Caddy and Csirke, 1983). This is because the fishing mortality on one species will not only affect the target species, but also cascade through the system by either increase the lower trophic levels or decrease the higher trophic levels (see section 3 and Chapter 5, Figure 5.6). The proportion of the total generated production that can be considered as surplus, that is the part which is not used to maintain the population at a given level, is extremely difficult to define, and in fish stock assessments mostly depends on the mathematical model chosen.

In ecology, the *ecotrophic coefficient* is defined as the proportion of the production over a period of time by trophic level (n) available as 'yield' (consumption) to the next trophic level (n+1). Dickie (1972) deduced, based on theoretical considerations, that the ecotrophic coefficient in nature is unlikely to exceed a value of 0.5, meaning that a maximum of half of the production (*P*) is available as MSY. In fisheries theory the *exploitation rate* (E) is defined as:

$$E = \frac{C}{P} = \frac{F}{Z} \qquad (5)$$

In single-species models, where man is the only predator, the exploitation rate also has a general recommended maximum value of 0.5, but derived from the principle that F should be equal to M giving $Z=2M$. In a multispecies fishery situation, however, the ecotrophic coefficient is the fraction that should be *shared* between fishers and the fish predators, implying that the exploitation rate should be equal or less than the ecotrophic coefficient (Kolding, 1993b).

The impact of fishing on a fish community can now be illustrated by combining the fisheries and ecological concepts defined in sections 1 to 3. In summary: The yield or catch is a fraction of the production and defined as $C = F \cdot \overline{B}$ (section 1 eqn.1). From the *P/B* ratio, production can be defined as $P = ZB$ (which shows why $C/P = F/Z$ in eqn. 5). The Maximum Sustainable Yield (*MSY*), which is the carrying capacity of the next trophic level, has a theoretical maximum value of around half the total production, thus:

$$MSY_{prey} = K_{predator} \approx \frac{P_{prey}}{2} \approx \frac{Z_{prey} \cdot \overline{B}_{prey}}{2} \tag{6}$$

The answer to the question how to share the MSY depends on how one wants the fish community to be composed. In the absence of other information, a conservative exploitation rate of 0.5 on top-level predators and 0.25 on lower levels could be used, which means that man becomes the new top predator and otherwise share the rest fifty-fifty. Such a fishing pattern will in theory keep the relative abundance of fish in the community unaltered, but will lower the overall biomass. This principle, together with the impact of different fishing patterns, is illustrated in Figure 2, which for simplicity, assumes a steady state community under logistic conditions (eqn 2) where $MSY = 0.5BZ$ at $B\infty/2$. The system is closed, the primary production finite, and we require that the original species composition (in this case 3 stocks at 3 trophic levels) should be conserved. Under unfished 'virgin' conditions (Figure 2A) the energy source of each trophic level is defined by the maximum 'yield' from the level below. We then start exploiting the system from the top level (Figure2B) by harvesting the MSY (i.e. reducing the 'virgin' biomass by half). This will decrease consumption by half and thus release half the 'yield' from the second level (MSY/2) for human exploitation, but in theory no 'surplus' is made available from the first level. In Figure 2C, exploitation starts from the bottom level. Removing a proportion of the MSY from the first level will reduce the 'carrying capacity' of the next level and thus reduce its 'virgin' steady state biomass to a new value: $B\infty^*$ ($B\infty^* < B\infty$). This reduction will cascade up through the system and also affect the potential yields ($MSY_n^* < MSY_n$), but in theory the system will find a new balance under the new carrying capacities. As the lower trophic levels are having the highest productivity (highest P/B ratio), the fishing pattern sketched in Figure 2C (equivalent to "hunting all that moves" in Figure 5.5) seems the most rational solution to exploiting the whole system (i.e. maximizing the output) without causing deep disturbances (Kolding, 1994). In theory, due to gear selectivity as described above, such an exploitation pattern in a multispecies community can only be achieved by employing a multitude of fishing gears, which in combination can generate a size-specific fishing mortality that is proportional to the natural size-specific mortality pattern. Incidentally, in contrast with most fisheries theory based on single-species considerations, effort in such a fishing pattern should in most systems be highest on the smaller sizes to match the prevailing natural mortality pattern (see Figure 1).

A) The 'virgin' unexploited fish community

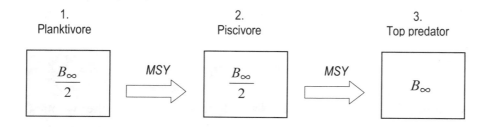

B) Exploitation of the community beginning from the top level

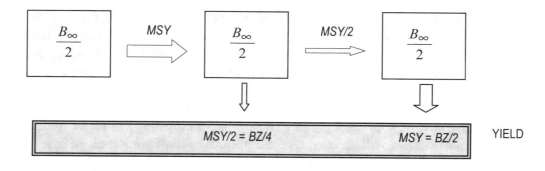

C) Exploitation of the community beginning from the bottom level

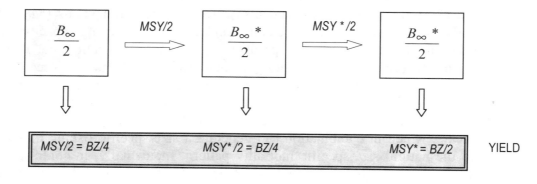

Figure 2 *A simplified fish community of three trophic levels. Each box represents the biomass of each level (not to scale; see Figure 5.5) relative to the 'virgin' biomass (B_∞) of each level, under logistic conditions. Arrows indicate the flow of energy (net production) through the system. See text for explanation. Redrawn from Kolding (1994)*

Section 5. Scale of operation and variability in daily catch rates

Variability in the day-to-day catch of a fisherman depends in the first place on the fishing pattern. The type of gear used is strongly related to its selectivity, while the scale at which it is operated determines both the magnitude of the effect it has on the stock and the variability in the day-to-day catch. Increasing the fished space – larger or more gillnets, hauls of longer duration in trawls – and increasing efficiency in finding fish by auxiliary devices will decrease variability. The first aspect in fact is simply an effect of aggregation of fish over space. The last aspect, the probability of encountering fish, in the first place depends on investments in mobility, that is boats. Increased mobility with paddles, sails or engines will increase the spatial allocation of effort or the freedom of movement to choose between fishing grounds and the operation of the gear. The possibility of increasing the encounter rate with fish (+ efficiency), and decreasing variability, also depends both on the knowledge a fisherman has about the spatial distribution of the fish (local knowledge) and on his means to invest in auxiliary technology, such as fish-finding devices.

Besides the choice of fishing gear, variability in catch also depends on the resource character. In Chapter 5 and in the previous sections of this Appendix resource character has been discussed extensively in the light of susceptibility of species and fish communities to fishing under annual flood pulses. What was not discussed was that patterns of spatial distribution and dispersal typical for a species or even a particular fish community, will determine the probability of encounter. Catches of schooling and aggregated species with low stock densities, distributed and moving around over large areas, for example many pelagics, are more variable compared to catches of species that are more evenly distributed over space with limited movement and/or with high stock densities (Densen, 2001). Lastly variability in day-to-day catches will increase with decreasing stock size (Figure 3 indicated by "- stock size").

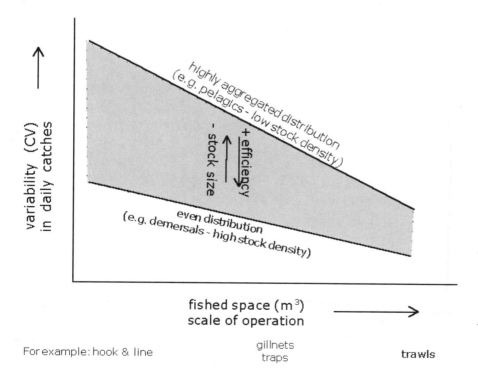

Figure 3 *Conceptual model classifying fishing methods based on their scale of operation and the resulting variability in catch probability (CV). The grey area indicates the range of daily variability of a particular gear. Efficiency includes both technological development of gears and fishermen's knowledge of the spatial behaviour of the fish. Decreasing stock size is a result of fishing, environment or migration.*